INTRODUCTION TO AFFINE ALGEBRAIC GROUPS

INTRODUCTION TO AFFINE ALGEBRAIC GROUPS

G. HOCHSCHILD

Department of Mathematics
University of California, Berkeley

HOLDEN-DAY, INC.

San Francisco, Cambridge, London, Amsterdam

500 Sansome Street, San Francisco, California.

ISBN 0-8162-3954-1

Library of Congress Catalog Card Number: 78-141286

Printed in the United States of America.

PREFACE

The theory of affine algebraic groups was originally developed in connection with Lie group theory, where these groups appeared as algebraic linear groups, i.e. as groups of automorphisms of finite-dimensional vector spaces, defined as subgroups of the full linear groups by algebraic conditions, such as the condition of fixing a given bilinear form. The first systematic and purely algebraic development of algebraic linear groups was carried out by Chevalley, in [2].* Concurrently with new developments in algebraic geometry, this theory was then built up to a theory of general algebraic groups, which contains the theory of algebraic linear groups as the special case of affine algebraic groups, i.e. groups whose underlying algebraic varieties are affine. Simultaneously, the structure theory of these affine groups was deepened and freed from the classical restriction to base fields of characteristic 0. A culmination of this development is [3]. Although the groups being investigated are affine, it soon becomes necessary to invoke much more than just affine algebraic geometry. The reason for this is that algebraic homogeneous spaces that are not *affine* algebraic varieties play a decisive role in the deeper structure theory of affine algebraic groups. A detailed account, *ab initio*, of the structure theory would therefore be excessively voluminous. For the student who is already equipped with algebraic geometry, the approach has recently been paved by Borel, in [1].

The present exposition has more modest aims. It is oriented toward representation theory, and its principal aim is to fuse representation-theoretical technique with the elementary theory of affine algebraic groups so as to provide an efficient tool, especially for Lie group theory.

* Numbers in brackets refer to the list of Supplementary Readings, to be found at the end of the book.

PREFACE

The main material covered is in essence, though not in form, that of [2]. Considerable emphasis is placed on the linearization technique based on Lie algebras, which becomes effective only when the base field is assumed to be of characteristic 0.

The setting in which we operate arises as follows. A finite-dimensional representation theory for any group G is based on the choice of a category of admissible G-modules, and hence on the choice of an associated family of representative functions on G. The family of these functions has the structure of a Hopf algebra over the chosen base field. The comultiplication of this Hopf algebra reflects the group multiplication of G, while the multiplication of the Hopf algebra (which is simply the value-wise multiplication of functions) reflects the diagonal map $G \to G \times G$, and corresponds to the formation of tensor products of G-modules. The G-modules under investigation may be viewed as the comodules for the Hopf algebra of representative functions. On the other hand, this Hopf algebra defines a group G^*: namely, the group of all finite specializations of the Hopf algebra into the base field. This group G^* may be regarded as a universal algebraic hull of G. It has the structure of a pro-affine algebraic group (i.e. a projective limit of affine algebraic groups), and the representation theory of G becomes the "rational" representation theory of G^* as a pro-affine algebraic group. The importance of the theory of affine algebraic groups for general representation theory is immediately evident from this. The passage from affine algebraic groups to pro-affine algebraic groups is simply the passage from finitely generated Hopf algebras to general ones, and every Hopf algebra is the union of the family of its finitely generated Hopf subalgebras. Once the affine theory has been brought under control, the extension to the pro-affine theory is not difficult. Here, we remain with the affine situation.

In order to keep this book self-contained, we have limited the material so that only elementary affine algebraic geometry comes into play, and all the results in this area that we use are established in Section 1. All that is presupposed with regard to the elementary theory of Lie algebras, used in the later sections, will be found in Chapters X and XI of [4] or, of course, in [5]. Otherwise, no special knowledge is presupposed, although it is assumed that the reader will be able to assimilate the daily diet of the working algebraist in unpremasticated form. The organization should be clear from the section headings.

G. H.

Berkeley, September, 1970

CONTENTS

1. ALGEBRAIC PRELIMINARIES

We begin by assembling a few basic facts concerning homomorphisms of rings into algebraically closed fields. All the rings we shall consider are assumed to be commutative and to have an identity element. The statement that B is a subring of A is to imply that the identity element of B coincides with that of A. The statement that ρ is a ring homomorphism of R into S is to imply that ρ sends the identity element of R onto that of S.

LEMMA 1.1. *Let R be a subring of a field K, and let J be an ideal of R, other than R. Then, for every element u of K, either $R[u]J \neq R[u]$ or $R[u^{-1}]J \neq R[u^{-1}]$.*

Proof: If this is false, then we have relations

$$\sum_{i=0}^{m} a_i u^i = 1 = \sum_{j=0}^{n} b_j u^{-j},$$

with the a_i's and b_j's in J. We assume that m and n are chosen as small as possible. Replacing u with u^{-1}, if necessary, we arrange to have n no greater than m. Since $J \neq R$, m and n must be positive. From the second relation, we obtain

$$(1 - b_0)u^m = \sum_{j=1}^{n} b_j u^{m-j}.$$

Multiplying the first relation by $1 - b_0$ and then substituting for $(1 - b_0)u^m$, we obtain

$$1 - b_0 = \sum_{i=0}^{m-1} (1 - b_0)a_i u^i + a_m \sum_{j=1}^{n} b_j u^{m-j}.$$

This may be written in the form $\sum_{p=0}^{m-1} c_p u^p = 1$, with each c_p in J. Thus we have a contradiction to the minimality of m, so that the lemma is established.

A subring S of a field K is called a *valuation subring* if, for every element u of K, either u or u^{-1} belongs to S.

PROPOSITION 1.2. *Let R be a subring of a field K, and let ρ be a ring homomorphism of R into an algebraically closed field F. Then ρ can be extended to a ring homomorphism of a valuation subring S of K into F, where $R \subset S$.*

Proof: An evident application of Zorn's lemma shows that, among the subrings T of K that contain R and are such that ρ can be extended to a ring homomorphism $T \to F$, there is a maximal one. Hence we may assume that R is already maximal, and prove that R is a valuation subring of K.

Let J denote the kernel of ρ. Since $\rho(1)$ is the identity element of F, we have $J \neq R$. Let u be any element of K. We must show that one of u or u^{-1} belongs to R. By Lemma 1.1, we may suppose that $R[u]J \neq R[u]$, and then it suffices to show that u belongs to R. Our last assumption implies that J is contained in some maximal ideal M of $R[u]$. Since ρ can evidently be extended to the ring of fractions a/b, with a and b in R, and b not in J, the maximality of R implies that R coincides with this ring of fractions. Hence $\rho(R)$ is a subfield of F, so that J is a maximal ideal of R. Hence we must have $J = M \cap R$.

Now let π denote the canonical homomorphism $R[u] \to R[u]/M$. Since the kernel of π in R is J, there is an isomorphism $\sigma : \pi(R) \to \rho(R)$ such that the restriction to R of $\sigma \circ \pi$ coincides with ρ. Next, let us observe that u must be algebraic over R, because otherwise ρ could evidently be extended to a homomorphism $R[u] \to F$. Hence $\pi(u)$ is algebraic over the subfield $\pi(R)$ of $R[u]/M$. Since F is algebraically closed, the isomorphism σ can therefore be extended to a homomorphism $\tau: R[u]/M \to F$. Now the homomorphism $\tau \circ \pi: R[u] \to F$ is an extension of ρ, so that the maximality of R implies that u belongs to R. This completes the proof of Proposition 1.2.

THEOREM 1.3. *Let R be a subring of a field K, and let P be a finite subset of K. Let u be a non-zero element of the subring $R[P]$ of K. Then there is a non-zero element u' in R such that every homomorphism of R into an algebraically closed field F that does not annihilate u' extends to a homomorphism of $R[P]$ into F that does not annihilate u.*

Proof: An evident induction on the number of elements of P reduces the theorem to the case where P consists of a single element p. First, let us consider the case where p is not algebraic over the field of fractions $[R]$ of R. Write

$$u = r_0 + r_1 p + \cdots + r_n p^n,$$

with each r_i in R, and $r_n \neq 0$. Now let ρ be a homomorphism of R into an algebraically closed field F such that $\rho(r_n) \neq 0$. Then there is an element t in F such that $\rho(r_0) + \cdots + \rho(r_n)t^n \neq 0$. Evidently, ρ can be extended to a homomorphism $\sigma : R[p] \to F$ such that $\sigma(p) = t$. Our choice of t ensures that $\sigma(u) \neq 0$, so that this case of the theorem is proved, with $u' = r_n$.

Now suppose that p is algebraic over $[R]$. Then we can find a non-zero element u' in R such that $u'p$ and $u'u^{-1}$ are integral over R, whence p and u^{-1} are integral over $R[u'^{-1}]$. Let ρ be a homomorphism of R into an algebraically closed field F such that $\rho(u') \neq 0$. Clearly, ρ can be extended to a homomorphism $\sigma : R[u'^{-1}] \to F$. By Proposition 1.2, there is a valuation subring S of K such that $R[u'^{-1}] \subset S$ and σ extends to a homomorphism $\tau : S \to F$. Since S is a valuation subring of K, it is integrally closed in K, so that p and u^{-1} belong to S. The restriction of τ to $R[p]$ is an extension of ρ, and we have $\tau(u) \neq 0$, because $\tau(u^{-1})$ is defined and $\tau(u)\tau(u^{-1}) = 1$. This completes the proof of Theorem 1.3.

LEMMA 1.4. *Let B be a commutative ring. Then the intersection of the family of all prime ideals of B coincides with the set of all nilpotent elements of B.*

Proof: Evidently, every nilpotent element of B must belong to every prime ideal of B. Conversely, suppose that b is an element of B that belongs to every prime ideal of B. Consider the polynomial ring $B[x]$ in one variable x, with coefficient ring B. Clearly, b, and therefore also bx, belongs to every maximal ideal of $B[x]$. Hence $1 - bx$ is a unit of $B[x]$, i.e., there are elements b_i in B such that

$$(1 + b_1 x + \cdots + b_n x^n)(1 - bx) = 1 .$$

One reads off from this that $b_1 = b$, $b_2 = b_1 b, \ldots, b_n = b_{n-1} b$, $b_n b = 0$, whence $b^{n+1} = 0$. This proves Lemma 1.4.

THEOREM 1.5. *Let L be a field, and let B be a finitely generated commutative L-algebra with identity element. Let F be an algebraically closed field containing L, and suppose that b is an element of B that is annihilated by every L-algebra homomorphism $B \to F$. Then b is nilpotent.*

Proof: Let J be a prime ideal of B. Then B/J is an integral domain and a finitely generated L-algebra (we identify L with its image in B/J). We may apply Theorem 1.3, with L in place of R and B/J in place of $R[P]$. This shows that if the image of b in B/J is not 0, there is an L-algebra homomorphism $B/J \rightarrow F$ that does not annihilate the image of b. The composite of this with the canonical homomorphism $B \rightarrow B/J$ is an L-algebra homomorphism $B \rightarrow F$ which does not annihilate b, contrary to assumption. Thus we must have $b \in J$. Now it follows from Lemma 1.4 that b is nilpotent, so that Theorem 1.5 is proved.

The reader may recognize Theorem 1.5 as one of the many versions of the Hilbert Nullstellensatz. The next theorem answers a more elaborate question concerning separation properties of homomorphisms.

THEOREM 1.6. *Let L be a field, and let F be an algebraically closed field containing L. Let A and B be finitely generated L-algebras that are integral domains, and suppose that $B \subset A$. Let x be an element of A with the following property: if ρ and σ are L-algebra homomorphisms $A \rightarrow F$ whose restrictions to B coincide, then $\rho(x) = \sigma(x)$. Then x is purely inseparably algebraic over the field of fractions of B.*

Proof: First, we show that x is algebraic over the field of fractions $[B]$ of B. Suppose this false. By Theorem 1.3, there is a non-zero element y in $B[x]$ such that every L-algebra homomorphism $\rho : B[x] \rightarrow F$ for which $\rho(y) \neq 0$ extends to an L-algebra homomorphism $A \rightarrow F$. Write

$$y = b_0 + b_1x + \cdots + b_nx^n,$$

with each b_i in B, and $b_n \neq 0$. Since B is an integral domain, it follows from Theorem 1.5 that there is an L-algebra homomorphism $\sigma : B \rightarrow F$ such that $\sigma(b_n) \neq 0$. Since x is not algebraic over $[B]$, and since F is infinite, σ has infinitely many extensions to L-algebra homomorphisms $B[x] \rightarrow F$ that do not annihilate y, and which therefore extend further to L-algebra homomorphisms $A \rightarrow F$. This contradicts the assumption on x, so that we may conclude that x is algebraic over $[B]$.

Let p stand for the characteristic of L if that is not 0, and let $p = 1$ if L is of characteristic 0. Suppose that x is not purely inseparable over $[B]$. Then there is an exponent e such that x^{p^e} is separably algebraic over $[B]$, but does not belong to $[B]$. We can find a non-zero element b in B such that the monic minimum polynomial of bx^{p^e} relative to $[B]$ has all its coefficients in B. Write z for bx^{p^e}, and f for its monic minimum polynomial. The degree of f is greater than 1; call it m. As before, there is a non-zero element y in $B[z]$ such that every L-algebra homomorphism of $B[z]$ into F that does not annihilate y extends to an L-algebra homomorphism of A into F. We have

$$y = b_0 + b_1 z + \cdots + b_n z^n,$$

with each b_i in B, $b_n \neq 0$, and $n < m$. Let g be the polynomial whose coefficients are these b_i's. Then we can find polynomials u and v with coefficients in B such that $uf + vg$ is a non-zero element s of B. On the other hand, since f is a separable polynomial, its formal derivative f' is different from 0, so that we can find polynomials q and r with coefficients in B such that $qf + rf'$ is a non-zero element t of B.

By Theorem 1.5, there is an L-algebra homomorphism ρ of B into F such that $\rho(st) \neq 0$. Let $\rho(f)$, etc. denote the polynomials with coefficients in F that are obtained by applying ρ to the coefficients of f, etc. Then, since $\rho(s)$ and $\rho(t)$ are different from 0, the polynomials $\rho(f)$ and $\rho(g)$ are relatively prime, as well as the polynomials $\rho(f)$ and $\rho(f')$. Moreover, since f is monic, $\rho(f)$ is still of degree $m > 1$. Hence $\rho(f)$ has m distinct roots in F, and none of them is a root of $\rho(g)$. Using these roots as values of z, we obtain m different extensions of ρ to L-algebra homomorphisms $B[z] \to F$, none of which annihilates y, so that each extends further to an L-algebra homomorphism $A \to F$. The images of x under these homomorphisms are all distinct, so that we have a contradiction to the assumption on x. Our proof of Theorem 1.6 is now complete.

Next, we shall deal briefly with chains of prime ideals, establishing only the bare minimum of basic results. We denote by $R[x]$ the polynomial ring in one variable x over the commutative ring R.

LEMMA 1.7. *Let P_0, P_1, P_2 be prime ideals of the polynomial ring $R[x]$ such that, with proper inclusions, $P_0 \subset P_1 \subset P_2$. Then $P_0 \cap R \neq P_2 \cap R$.*

Proof: Let Q_0 denote the ideal of $R[x]$ that is generated by $P_0 \cap R$. Evidently, Q_0 is a prime ideal and is contained in P_0. We identify $R[x]/Q_0$ with the polynomial ring $R'[x]$, where R' is the integral domain $R/(P_0 \cap R)$. Now P_1/Q_0 and P_2/Q_0 are prime ideals of $R'[x]$, and we have the proper inclusions $(0) \subset P_1/Q_0 \subset P_2/Q_0$. Forming tensor products relative to R', we obtain ideals $[R'] \otimes (P_1/Q_0)$ and $[R'] \otimes (P_2/Q_0)$ of the polynomial ring $[R'][x]$. The first of these is not (0) and is contained in the second. Each of them is either a prime ideal or coincides with $[R'][x]$. Since every non-zero prime ideal of $[R'][x]$ is maximal, it follows that we must have either

$$[R'] \otimes (P_1/Q_0) = [R'] \otimes (P_2/Q_0) \quad \text{or} \quad [R'] \otimes (P_2/Q_0) = [R'][x] .$$

The second alternative means that P_2/Q_0 contains a non-zero element of R'. A representative in $P_2 \cap R$ of this element does not belong to P_0, so that we have $P_0 \cap R \neq P_2 \cap R$, in this case.

Now consider the other alternative $[R'] \otimes (P_1/Q_0) = [R'] \otimes (P_2/Q_0)$. Choose an element p of P_2 that does not belong to P_1. The present assumption implies that there is a non-zero element α of R' such that the product of α and the image of p in P_2/Q_0 belongs to P_1/Q_0. Let a be a representative of α in R. Then we have $ap \in P_1$. Since p does not belong to P_1, it follows that $a \in P_1$. Since α is not 0, the representative a does not belong to P_0. Hence $P_1 \cap R \neq P_0 \cap R$. *A fortiori*, $P_2 \cap R \neq P_0 \cap R$, so that Lemma 1.7 is established.

If $P_0 \subset P_1 \subset \cdots \subset P_n$ is a chain of prime ideals of the commutative ring R, and if the inclusions are proper, then n is called the length of the chain. If there is a greatest possible length for such chains, this length is called the *Krull dimension* of R, and we say then that R has finite Krull dimension.

PROPOSITION 1.8. *Let F be a field, and let R be a finitely generated F-algebra that is an integral domain. Then R has finite Krull dimension.*

Proof: We write $R = F[u_1, \ldots, u_n]$, and we make an induction on the number n of F-algebra generators. If $n = 0$, then $R = F$, and its Krull dimension is evidently 0. Now it suffices to show that if S is any integral domain of finite Krull dimension, and if $R = S[u]$ is an integral domain containing S and being generated, as an S-algebra, by a single element u, then R has finite Krull dimension.

Let x be an auxiliary variable, and write $S[u]$ as the factor algebra of the polynomial algebra $S[x]$ modulo a prime ideal P. It is clear from Lemma 1.7 that $S[x]$ still has finite Krull dimension. The prime ideals of $S[u]$ are of the form Q/P, where Q is a prime ideal of $S[x]$ containing P. Hence it is clear that $S[u]$ has finite Krull dimension, which is at most equal to that of $S[x]$. This proves Proposition 1.8.

The next three results are needed in order to establish Theorem 1.12, which is important for our main purpose.

LEMMA 1.9. *Let F be a field, and let $A = F[a_1, \ldots, a_n]$ be a finitely generated commutative F-algebra. Let I be an ideal of A, with $I \neq A$. Then there are elements $b_1, \ldots, b_n$ in A, and an index $r \leqq n$, such that*

$b_i \in I$ for each $i < r$, $I \cap F[b_r, \ldots, b_n] = (0)$, and A is integral over $F[b_1, \ldots, b_n]$.

Proof: We make an induction on the number n of generators of A. If $n = 0$ we have $A = F$, $I = (0)$, and the lemma holds trivially. Suppose that $n > 0$, and that the lemma has been established for algebras with fewer than n generators. If $I = (0)$, we may evidently take $r = 1$ and each $b_i = a_i$. Hence we may assume that I is not (0).

Let b_1 be any non-zero element of I, and write $b_1 = f(a_1, \ldots, a_n)$, where f is a polynomial in n variables $x_1, \ldots, x_n$, with coefficients in F. Let d denote the total degree of f. Since $I \neq A$, we must have $d > 0$. For each $i > 1$, put

$$c_i = a_i - a_1^{(d+1)^{i-1}}.$$

Then we have

$$b_1 = f(a_1, c_2 + a_1^{d+1}, \ldots, c_n + a_1^{(d+1)^{n-1}}).$$

The term of highest degree in x_1 in the expansion of

$$x_1^{e_1}(x_2 + x_1^{d+1})^{e_2} \cdots (x_n + x_1^{(d+1)^{n-1}})^{e_n}$$

is

$$x_1^{e_1 + e_2(d+1) + \cdots + e_n(d+1)^{n-1}}.$$

As the e_i's range over the natural numbers from 0 to d, the exponent of x_1 here is never repeated. It follows that, if we arrange the full expansion of our above expression for b_1 according to the powers of a_1, we obtain

$$b_1 = ua_1^e + \sum_{i=0}^{e-1} f_i(c_2, \ldots, c_n)a_1^i,$$

where u is a non-zero element of F, the exponent e is positive, and the f_i's are polynomials with coefficients in F. This shows that a_1 is integral over $F[b_1, c_2, \ldots, c_n]$. From the definition of the c_i's, it is clear that therefore each a_i is integral over $F[b_1, c_2, \ldots, c_n]$.

Now let $B = F[c_2, \ldots, c_n]$, and $J = I \cap B$. By our inductive hypothesis, there are elements $b_2, \ldots, b_n$ in B, and an index $r \leqq n$, such that $b_i \in J$ for each $i < r$, $J \cap F[b_r, \ldots, b_n] = (0)$, and B is integral over $F[b_2, \ldots, b_n]$. Clearly, we have $b_i \in I$ for each $i < r$, and

$$I \cap F[b_r, \ldots, b_n] = (0).$$

Finally, since each a_i is integral over $F[b_1, c_2, \ldots, c_n]$, while $F[b_1, c_2, \ldots, c_n]$ is integral over $F[b_1, \ldots, b_n]$, it follows that each a_i is integral over $F[b_1, \ldots, b_n]$. This completes the proof of Lemma 1.9.

Now we can readily deduce Noether's Normalization Theorem, which is as follows.

THEOREM 1.10. *Let F be a field, and let* $R = F[r_1, \ldots, r_n]$ *be a finitely generated commutative F-algebra. Then there is a set* $(z_1, \ldots, z_s)$ *of elements of R (where* $s \leqq n$*) that is algebraically free over F and such that R is integral over the polynomial algebra* $F[z_1, \ldots, z_s]$.

Proof: Let $x_1, \ldots, x_n$ be auxiliary variables over F, and write R as a factor algebra of the polynomial algebra $F[x_1, \ldots, x_n]$ modulo an ideal I. Let r, and $(b_1, \ldots, b_n)$, be as obtained in Lemma 1.9, with $F[x_1, \ldots, x_n]$ for A. Let z_i denote the canonical image in R of b_{i+r-1} $(i = 1, \ldots, n + 1 - r)$.

Since $F[x_1, \ldots, x_n]$ is integral over $F[b_1, \ldots, b_n]$, we have R integral over the image of $F[b_1, \ldots, b_n]$ in R. This image is $F[z_1, \ldots, z_{n+1-r}]$, because b_i belongs to I for each $i < r$. It is clear that $(b_1, \ldots, b_n)$ must be algebraically free over F. Since $I \cap F[b_r, \ldots, b_n] = (0)$, this implies that $(z_1, \ldots, z_{n+1-r})$ is algebraically free over F. Thus Theorem 1.10 is established.

LEMMA 1.11. *Let A be a Noetherian integral domain that is integrally closed in its field of fractions* $[A]$. *Let L be a finite separable algebraic extension field of* $[A]$. *Then the integral closure of A in L is Noetherian as an A-module.*

Proof: Let T denote the trace map $L \to [A]$. Since L is separable over $[A]$, the $[A]$-bilinear trace form $(u, v) \to T(uv)$ on $L \times L$ is non-degenerate. Let A^L denote the integral closure of A in L. Clearly, A^L contains an $[A]$-basis $(u_1, \ldots, u_n)$ of L. Because of the non-degeneracy of the trace form, we can find elements $t_1, \ldots, t_n$ in L such that $T(u_i t_j)$ is equal to 1 or 0 according to whether $i = j$ or $i \neq j$. Now let u be any element of A^L. Evidently, $(t_1, \ldots, t_n)$ is also an $[A]$-basis of L, so that we have $u = \sum_{i=1}^{n} s_i t_i$, with each s_i in $[A]$. Multiplying by u_j and then taking the trace, we find that $s_j = T(u_j u)$. This is integral over A, because $u_j u$ belongs to A^L. Since A is integrally closed in $[A]$, we have therefore $s_j \in A$. Thus A^L is contained in the finitely generated A-module $At_1 + \cdots + At_n$. Since A is Noetherian, it follows that A^L is Noetherian as an A-module, so that Lemma 1.11 is proved.

THEOREM 1.12. *Let A be an integral domain containing the field F and being finitely generated as an F-algebra. Let L be a finite algebraic extension field of the field of fractions* $[A]$. *Then the integral closure of A in L is a Noetherian A-module.*

Proof: By Theorem 1.10, there is a finite subset $(z_1, \ldots, z_s)$ of A that is algebraically free over F and such that A is integral over the polynomial algebra $B = F[z_1, \ldots, z_s]$. Clearly, the integral closure of B in L coincides with the integral closure of A. If this integral closure is Noetherian as a B-module than, *a fortiori*, it is finitely generated as an A-module, and hence Noetherian as an A-module, because A is Noetherian. Hence it suffices to prove the theorem for B, i.e., we may assume that A is an ordinary polynomial algebra $F[x_1, \ldots, x_n]$ with $(x_1, \ldots, x_n)$ algebraically free over F.

Let L° be an algebraic closure of L. Then L° contains a purely inseparable algebraic extension P of $[A]$ such that P is a perfect field. Let $(y_1, \ldots, y_m)$ be a set of field generators for L over $[A]$. The coefficients of the monic minimum polynomials of the y_i's with respect to P generate a finite purely inseparable algebraic extension field S of $[A]$, and the subfield $S[L]$ of L° is a finite separable algebraic extension of S. We have $S = [A][u_1, \ldots, u_r]$, and there is a power q of the characteristic of F such that $u_j^q \in [A]$ for each j; in the case where F is of characteristic 0, we have $S = [A]$, and the present part of the proof becomes vacuous.

Now $[A] = F(x_1, \ldots, x_n)$, so that each u_j^q is a fraction formed with two polynomials in the x_i's, with coefficients in F. Let $a_1, \ldots, a_s$ be all these coefficients. Working in L°, put

$$B = A[x_1^{1/q}, \ldots, x_n^{1/q}, a_1^{1/q}, \ldots, a_s^{1/q}]\,.$$

Then B is integral over A. Let K be the extension field of F that is generated by the elements $a_i^{1/q}$. Then $(x_1^{1/q}, \ldots, x_n^{1/q})$ is algebraically free over K, and B is the polynomial algebra $K[x_1^{1/q}, \ldots, x_n^{1/q}]$. Hence B is integrally closed in its field of fractions

$$[B] = [A][x_1^{1/q}, \ldots, x_n^{1/q}, a_1^{1/q}, \ldots, a_s^{1/q}]\,.$$

which evidently contains S. The integral closure A^S of A in S is therefore contained in B. Since B is a finitely generated A-module, and since A is Noetherian, it follows that A^S is Noetherian as an A-module. *A fortiori*, A^S is a Noetherian ring. Moreover, A^S is integrally closed in its field of fractions, because this field of fractions is S.

Now $S[L]$ is a finite separable algebraic extension field of S, so that we may apply Lemma 1.11 to conclude that the integral closure of A^S in $S[L]$ is a Noetherian A^S-module. Since A^S is a Noetherian A-module, it follows that the integral closure of A^S in $S[L]$ is also a Noetherian A-module. But this integral closure evidently contains the integral closure A^L of A in L. Hence A^L is a Noetherian A-module, and Theorem 1.12 is proved.

We shall have occasion to use the following very effective and completely elementary result due to Artin and Tate, who have used it for giving a short proof of the Hilbert Nullstellensatz.

THEOREM 1.13. *Let A and B be commutative algebras over the Noetherian commutative ring R, with $R \subset B \subset A$. Suppose that A is finitely generated as an R-algebra, and also as a B-module. Then B is finitely generated as an R-algebra.*

Proof: Write $A = R[a_1, \ldots, a_n] = Bu_1 + \cdots + Bu_m$, choosing $u_1 = 1$. Then $a_i = \sum_{j=1}^m b_{ij}u_j$, with each b_{ij} in B. Also $u_iu_j = \sum_{k=1}^m b_{ijk}u_k$, with each b_{ijk} in B. Let C denote the R-subalgebra of B that is generated by all these coefficients b_{ij} and b_{ijk}. Since R is Noetherian and C is a finitely generated R-algebra, we have that C is a Noetherian ring. Clearly, $Cu_1 + \cdots + Cu_m$ is a subring of A, and it contains R as well as each a_i. Hence it coincides with A, so that A is a finitely generated C-module. Since C is Noetherian, the C-submodule B of A is also finitely generated; $B = Cb_1 + \cdots + Cb_q$. Thus the elements b_p, b_{ij}, and b_{ijk} generate B as an R-algebra, so that Theorem 1.13 is proved.

We finish off this section with a few basic results concerning linear endomorphisms of vector spaces. Recall that a module is said to be *semisimple* if every submodule is a direct module summand, and that this is so if and only if the module is a sum of simple submodules.

PROPOSITION 1.14. *Let V be a vector space over a field F, and let S be a set of F-linear endomorphisms of V. Let T be a field containing F. Then, if $V \otimes_F T$ is semisimple with respect to the canonical extension of S to a set of T-linear endomorphisms of $V \otimes_F T$, it follows that V is semisimple with respect to S. Conversely, if V is finite-dimensional, and semisimple with respect to S, and if F is a perfect field, then $V \otimes_F T$ is semisimple with respect to the canonical extension of S.*

Proof: Suppose that $V \otimes T$ is semisimple, and let U be any S-stable subspace of V. Then $U \otimes T$ is a T-subspace of $V \otimes T$ that is stable under the canonical extension of S. By assumption on $V \otimes T$, there is an S-module homomorphism μ of $(V \otimes T)/(U \otimes T)$ into $V \otimes T$ such that $\pi \circ \mu$ is the identity map on $(V \otimes T)/(U \otimes T)$, where π is the canonical map $V \otimes T \rightarrow (V \otimes T)/(U \otimes T)$. Identify $(V \otimes T)/(U \otimes T)$ with $(V/U) \otimes T$, and note that π becomes the canonical extension of the canonical map $V \rightarrow V/U$. Now let us choose an F-space complement T_1 of F in T. For x in V/U, let us write $\mu(x) = \mu_0(x) + \mu_1(x)$, with uniquely determined components $\mu_0(x)$ in V and $\mu_1(x)$ in $V \otimes T_1$.

This defines a map μ_0 of V/U into V, and it is clear that μ_0 is an S-module homomorphism whose composite with the canonical map $V \to V/U$ is the identity map on V/U. Hence we see that V is semisimple with respect to S.

In proving the converse, we may evidently assume that V is simple with respect to S. Moreover, if T' is an algebraic closure of T, and if we show that $V \otimes T'$ is semisimple, then it follows from what we have already proved that $V \otimes T$ is semisimple. Hence we may assume also that T is algebraically closed. Assuming this, let G be the group of all field automorphisms of T leaving the elements of F fixed. Since F is perfect, the G-fixed elements of T are precisely the elements of F.

Now G acts on $V \otimes T$ in the natural way by F-linear (and T-semilinear) automorphisms, and the G-operators on $V \otimes T$ commute with the endomorphisms coming from the elements of S. Hence the G-operators permute the S-stable T-subspaces of $V \otimes T$ among themselves. Since $V \otimes T$ is finite-dimensional as a T-space, there is a non-zero S-stable and S-simple T-subspace A of $V \otimes T$. Put

$$B = \sum_{g \in G} g(A) .$$

Then B is evidently an S-stable and S-semisimple T-subspace of $V \otimes T$, and it suffices to show that $B = V \otimes T$.

Clearly, B is G-stable. Consider the non-zero elements of B, writing them in the form $\sum_{i=1}^{n} t_i v_i$, where the t_i's are in T, and $v_1, \ldots, v_n$ are F-linearly independent (hence T-linearly independent) elements of V. Take such a sum in which n is minimal, and multiply by t_1^{-1}, if necessary, to insure $t_1 = 1$. Applying an element g of G to this sum, and then subtracting the original sum from the transform, we obtain a sum in B with fewer than n summands, so that we must obtain 0, i.e., we must have $g(t_i) = t_i$ for each i, and for every g in G. Hence each t_i belongs to F, so that our sum belongs to V. Thus we have $B \cap V \neq (0)$. Since V is S-simple, this implies that $B \cap V = V$, whence $B = V \otimes T$, *q.e.d.*

THEOREM 1.15. *Let V be a finite-dimensional vector space over a perfect field F, and let e be an F-linear endomorphism of V. There are linear endomorphisms $e^{(n)}$ and $e^{(s)}$ of V having the following properties: $e^{(n)}$ is nilpotent, $e^{(s)}$ is semisimple, $e^{(n)} + e^{(s)} = e$, and both $e^{(n)}$ and $e^{(s)}$ belong to $\sum_{i>0} Fe^i$. Moreover, if f and g are linear endomorphisms of V such that $fg = gf$, f is nilpotent, g is semisimple, and $f + g = e$, then $f = e^{(n)}$ and $g = e^{(s)}$.*

Proof: We obtain $e^{(s)}$ directly from the following ingenious construction due to Chevalley. Let $F[x]$ denote the polynomial algebra in one variable x over F, and let ρ be the F-algebra homomorphism of $F[x]$ into the algebra of all linear endomorphisms of V that sends x onto e. The kernel of ρ is a non-zero ideal of $F[x]$, which is generated by a non-zero element p of $F[x]$. If p is divisible by x we define q as the product of the distinct monic irreducible factors of p. If p is not divisible by x we define q as the product of x and the distinct monic irreducible factors of p. Then q is a polynomial with constants term 0, and there is a positive integer k such that $\rho(q^k) = 0$. Since F is perfect, the polynomial q has no multiple roots, so that q is relatively prime to its formal derivative q'. Thus there are elements u and v in $F[x]$ such that $uq' + vq = 1$.

Now let σ denote the F-algebra endomorphism of $F[x]$ such that $\sigma(x) = x - uq$. Choose a positive integer m such that $2^m \geqq k$, and let a denote the linear endomorphism $\rho(\sigma^m(x))$ of V. Evidently, a belongs to $\sum_{i>0} Fe^i$. It follows immediately from Taylor's formula that $\sigma(q) = q - q'uq + r(uq)^2$, with r some element of $F[x]$. Since $q - q'uq = (1 - q'u)q = vq^2$, this shows that $\sigma(q)$ lies in $F[x]q^2$. Inductively, we obtain from this that $\sigma^m(q) \in F[x]q^{2m} \subset F[x]q^k$, whence $\rho(\sigma^m(q)) = 0$. But $\rho \circ \sigma^m$ is the F-algebra homomorphism of $F[x]$ into the endomorphism algebra of V that sends x onto a. Hence our last result shows that the F-algebra $\sum_{i\geqq 0} Fa^i$ is a homomorphic image of $F[x]/F[x]q$. Since q has no multiple prime factors, this last F-algebra is a direct sum of extension fields of F, whence the same is true for its homomorphic image $\sum_{i\geqq 0} Fa^i$. It is clear from this that a must be a semisimple linear endomorphism of V.

Now observe that $x - \sigma(x) \in F[x]q$. Since $\sigma(q)$ belongs to $F[x]q$, repeated applications of σ to this result show that $x - \sigma^m(x) \in F[x]q$. Hence $\rho(x - \sigma^m(x))^k = 0$, i.e., $(e - a)^k = 0$. Hence we have the required decomposition $e = e^{(n)} + e^{(s)}$, with $e^{(s)} = a$.

Finally, let f and g be as described in the statement of the theorem. Evidently, f and g commute with e, and hence also with $e^{(n)}$ and $e^{(s)}$. It follows that $f - e^{(n)}$ is nilpotent, and that $e^{(s)} - g$ is semisimple. Since these two differences are equal to each other, each is both nilpotent and semisimple, and hence must be 0. This completes the proof of Theorem 1.15.

The decomposition of an endomorphism, as described in Theorem 1.15, is called the (additive) *Jordan decomposition*. The components $e^{(n)}$ and $e^{(s)}$ are called the nilpotent and the semisimple component of e, respectively. In the case where e is an *automorphism* of V, the additive

Jordan decomposition gives rise to a *multiplicative Jordan decomposition*, as follows. One says that a vector space automorphism f is *unipotent* if $1 - f$ is nilpotent, where 1 stands for the identity automorphism. Now suppose that the endomorphism e of Theorem 1.15 is an automorphism. We claim that its semisimple component $e^{(s)}$ is then also an automorphism. In order to prove this, let W denote the $e^{(s)}$-annihilated part of V. It suffices to show that $W = (0)$. Since $e^{(n)}$ commutes with $e^{(s)}$, we have $e^{(n)}(W) \subset W$. Since $e^{(n)}$ is nilpotent, it follows that if $W \neq (0)$ then it contains a non-zero element w such that $e^{(n)}(w) = 0$. But then $e(w) = 0$, contradicting the fact that e is an automorphism. Now define $e^{(u)} = 1 + (e^{(s)})^{-1}e^{(n)}$. One sees immediately that $e = e^{(s)}e^{(u)}$, that $e^{(u)}$ is unipotent, and that $e^{(s)}$ and $e^{(u)}$ commute with each other. As in the case of the additive Jordan decomposition, one shows readily that $(e^{(s)}, e^{(u)})$ is the only pair of automorphisms satisfying these conditions. Moreover, since $e^{(s)}$ is an automorphism, $(e^{(s)})^{-1}$ and the identity automorphism 1 both belong to $\sum_{i>0} F(e^{(s)})^i$. Hence we see that $e^{(u)}$ (like $e^{(s)}$) belongs to $\sum_{i>0} Fe^i$. The automorphism $e^{(u)}$ is called the *unipotent component* of e.

EXERCISES

1. Let R be a subring of a field K, and let u be a non-zero element of K. Show that u is integral over R if and only if $u^{-1}R[u^{-1}]$ contains 1. Use this and Proposition 1.2 to prove that the integral closure of R in K is the intersection of the family of all valuation subrings of K that contain R.

2. Let F be an algebraically closed field, and let $(p_1, \ldots, p_m)$ be an algebraically independent set of polynomials in n variables with coefficients in F. Use Theorem 1.3 to show that there is a non-zero polynomial f in m variables with coefficients in F such that, for every m-tuple $(b_1, \ldots, b_m)$ of elements of F for which $f(b_1, \ldots, b_m) \neq 0$, there is an n-tuple $(a_1, \ldots, a_n)$ of elements of F such that $p_i(a_1, \ldots, a_n) = b_i$ for each i.

3. Let R be a Noetherian commutative ring, and let A be a finitely generated commutative R-algebra. Let G be a finite group of R-algebra automorphisms of A. Use Theorem 1.13 to show that the G-fixed part of A is finitely generated as an R-algebra.

4. Let e be a linear endomorphism of a finite-dimensional vector space V over an algebraically closed field F. Describe the semisimple component $e^{(s)}$ explicitly with reference to the nilspace decomposition of V corresponding to the characteristic roots of e.

2. GROUP REPRESENTATIONS AND HOPF ALGEBRAS

Let F be a field, and let S and T be non-empty sets. Denote by $\mathscr{M}_F(S)$ the F-algebra of all F-valued functions on S. There is a canonical map $\pi : \mathscr{M}_F(S) \otimes \mathscr{M}_F(T) \to \mathscr{M}_F(S \times T)$, where

$$\pi(f \otimes g)(s, t) = f(s)g(t) .$$

Evidently, π is an F-algebra homomorphism, and it is an elementary exercise in linear alegebra to show that π is injective. Moreover, it is readily seen that the image of π consists precisely of those elements h of $\mathscr{M}_F(S \times T)$ for which the space of functions $S \to F$ that is spanned by the partial maps h_t, where $h_t(s) = h(s, t)$, is finite-dimensional. An equivalent condition is obtained, of course, by switching the roles of S and T.

Now suppose that $S = G = T$, where G is an arbitrary group. The group multiplication $G \times G \to G$ naturally induces an F-algebra homomorphism $\sigma : \mathscr{M}_F(G) \to \mathscr{M}_F(G \times G)$, where $\sigma(f)(x, y) = f(xy)$. We define the left and right *translates* $x \cdot f$ and $f \cdot x$ of f by the element x of G as the functions $G \to F$ given by $(x \cdot f)(y) = f(yx)$ and $(f \cdot x)(y) = f(xy)$. Then it is clear from the above that $\sigma(f)$ belongs to the image of $\pi : \mathscr{M}_F(G) \otimes \mathscr{M}_F(G) \to \mathscr{M}_F(G \times G)$ if and only if the space of functions spanned by the translates of f (from the left, or, equivalently, from the right) is finite-dimensional. A function with this property is called a *representative function* on G. Clearly, the representative functions constitute an F-subalgebra $\mathscr{R}_F(G)$ of $\mathscr{M}_F(G)$.

Now $\pi^{-1} \circ \sigma$ is defined precisely on $\mathscr{R}_F(G)$, and one sees immediately that its image actually lies in the subalgebra $\mathscr{R}_F(G) \otimes \mathscr{R}_F(G)$ of $\mathscr{M}_F(G) \otimes \mathscr{M}_F(G)$. Thus $\pi^{-1} \circ \sigma$ may be regarded as an F-algebra homomorphism $\gamma : \mathscr{R}_F(G) \to \mathscr{R}_F(G) \otimes \mathscr{R}_F(G)$. The F-algebra structure of the tensor product involves the usual switching of the middle tensor factors: $(u_1 \otimes u_2)(v_1 \otimes v_2) = (u_1 v_1) \otimes (u_2 v_2)$. Element-wise, the connection between γ and the group multiplication of G is seen by noting that if $\gamma(f) = \sum f_i \otimes f_j$ then $f(xy) = \sum f_i(x) f_j(y)$.

The associativity of the group multiplication of G implies that γ is *co-associative*, i.e., that

$$(i \otimes \gamma) \circ \gamma = (\gamma \otimes i) \circ \gamma ,$$

where i stands for the identity map on $\mathscr{R}_F(G)$.

The evaluation of $\mathscr{R}_F(G)$ at the neutral element of G, $f \to f(1_G)$, is the *co-unit* (often called the *augmentation*) c for the comultiplication γ, i.e.,

$$(c \otimes i) \circ \gamma = i = (i \otimes c) \circ \gamma .$$

The structure $\big(\mathscr{R}_F(G), \gamma, c\big)$ is that of a *co-algebra*, with *comultiplication* γ, and co-unit c. The general notion of a co-algebra is obtained by replacing $\mathscr{R}_F(G)$ with an arbitrary F-space and demanding that γ and c be F-linear maps satisfying the formal conditions exhibited above.

Let U be a finite-dimensional G-module over F, and let ρ denote the representation of G on U, viewed as a group homomorphism of G into the group of units of the F-algebra $\mathscr{E}(U)$ of all F-space endomorphisms of U. Let $\mathscr{E}(U)^\circ$ denote the dual space of $\mathscr{E}(U)$. The composites $\tau \circ \rho$, where τ ranges over $\mathscr{E}(U)^\circ$, constitute an F-subspace of $\mathscr{R}_F(G)$, which we shall denote by $\mathscr{S}(U)$, or also by $\mathscr{S}(\rho)$, and which we call the space of *representative functions associated with* U, or with ρ. More generally, let V be a *locally finite* G-module, i.e. a G-module that coincides with the union of the family of its finite-dimensional G-submodules. Then we define $\mathscr{S}(V)$ as the union of the family of the subspaces $\mathscr{S}(U)$ of $\mathscr{R}_F(G)$, with U ranging over the family of finite-dimensional G-submodules of V. Clearly, $\mathscr{S}(V)$ is an F-subspace of $\mathscr{R}_F(G)$.

Conversely, we may regard $\mathscr{R}_F(G)$ as a locally finite G-module, the representation ρ of G on $\mathscr{R}_F(G)$ being given by $\rho(x)(f) = x \cdot f$. We have then $\mathscr{S}\big(\mathscr{R}_F(G)\big) = \mathscr{R}_F(G)$. In order to see this, let $[f]$ denote the smallest G-submodule of $\mathscr{R}_F(G)$ that contains a given element f of $\mathscr{R}_F(G)$. Let c/f denote the element of $\mathscr{E}([f])^\circ$ that is defined by $(c/f)(e) = c\big(e(f)\big)$ for every linear endomorphism e of $[f]$. Then $(c/f) \circ \rho = f$, showing that $f \in \mathscr{S}([f]) \subset \mathscr{S}\big(\mathscr{R}_F(G)\big)$.

Let V be an arbitrary F-space. As an evident generalization of representative functions, we may consider the *representative maps* of G into V, i.e. the maps $f : G \to V$ whose translates (defined as for F-valued functions) span only a finite-dimensional space of maps. If $\mathscr{R}_V(G)$ denotes the F-space of representative maps $G \to V$, then we may identify $\mathscr{R}_V(G)$ with the tensor product $V \otimes \mathscr{R}_F(G)$ by means of the evident map $V \otimes \mathscr{R}_F(G) \to \mathscr{R}_V(G)$, which sends $v \otimes f$ onto the map $G \to V$ whose value at x is $f(x)v$. In fact, this map is evidently injective, and we show that it is also surjective, as follows. Let f be an element of $\mathscr{R}_V(G)$, and let $(f_1, \ldots, f_n)$ be an F-basis for the space spanned by the translates $x \cdot f$. Then there are F-valued functions g_i on G such that $x \cdot f = \sum_{i=1}^n g_i(x) f_i$ for every x in G. Evaluating at the neutral element of G, we obtain $f = \sum_{i=1}^n c(f_i) g_i$, and it will suffice to show that each g_i belongs to $\mathscr{R}_F(G)$. This is seen upon writing

$$\sum_{i=1}^n g_i(xy) f_i = (xy) \cdot f = x \cdot (y \cdot f) = \sum_{i=1}^n g_i(y)(x \cdot f_i)$$
$$= \sum_{i,j} g_i(y) g_{ij}(x) f_j ,$$

where the g_{ij}'s are certain F-valued functions on G, which shows that

$$y \cdot g_j = \sum_{i=1}^n g_i(y) g_{ij} .$$

Now suppose that V is a locally finite G-module, and let ρ denote the representation of G on V. For every element v of V, let $\rho^*(v)$ denote the map of G into V that is given by $\rho^*(v)(x) = \rho(x)(v)$. From the fact that v belongs to some *finite-dimensional* G-submodule of V, one has that $\rho^*(v)$ belongs to $\mathscr{R}_V(G) = V \otimes \mathscr{R}_F(G)$. Moreover, $\rho^*(v)$ actually belongs to the subspace $V \otimes \mathscr{S}(V)$ of $V \otimes \mathscr{R}_F(G)$, as is readily seen by considering the composites of $\rho^*(v)$ with linear functionals on V. Clearly, ρ^* is a linear map $V \to V \otimes \mathscr{R}_F(G)$.

For any set T, let i_T denote the identity map on T. The fact that $\rho(xy) = \rho(x)\rho(y)$ for all elements x and y of G translates into the formula

$$(\rho^* \otimes i_{\mathscr{R}_F(G)}) \circ \rho^* = (i_V \otimes \gamma) \circ \rho^* ,$$

and the fact that $\rho(1_G) = i_V$ translates into the formula

$$(i_V \otimes c) \circ \rho^* = i_V .$$

Together, these properties of ρ^* mean than $\rho^* : V \to V \otimes \mathscr{R}_F(G)$ is the structure of a *comodule* for the co-algebra $\mathscr{R}_F(G)$. Observe that $\gamma : \mathscr{R}_F(G) \to \mathscr{R}_F(G) \otimes \mathscr{R}_F(G)$ is an example of a comodule structure; it corresponds to the locally finite G-module structure of $\mathscr{R}_F(G)$ discussed above.

Conversely, let $\tau : V \to V \otimes \mathscr{R}_F(G)$ be a comodule structure. For every element x of G, let x° stand for the evaluation of $\mathscr{R}_F(G)$ at x. Define the linear endomorphism (actually, an automorphism) $\tau_*(x)$ of V by

$$\tau_*(x) = (i_V \otimes x^\circ) \circ \tau .$$

It is easy to verify that τ_* is a representation of G on V, through which V becomes a locally finite G-module. Moreover, a straightforward computation shows that $(\tau_*)^* = \tau$, and $(\rho^*)_* = \rho$. Thus the correspondences $\rho \to \rho^*$ and $\tau \to \tau_*$ are mutually inverse, and they establish an isomorphism of the category of locally finite G-modules onto the category of $\mathscr{R}_F(G)$-comodules.

The algebra structure of $\mathscr{R}_F(G)$ enters into the G-module theory only when one considers tensor products of G-modules. This connection becomes transparent in terms of comodules, as follows. Let us write A for $\mathscr{R}_F(G)$, and let $\sigma : U \to U \otimes A$ and $\tau : V \to V \otimes A$ be two comodule structures. Then we have the maps

$$\begin{aligned} U \otimes V &\to (U \otimes A) \otimes (V \otimes A) \\ &\approx (U \otimes V) \otimes (A \otimes A) \to (U \otimes V) \otimes A , \end{aligned}$$

the first map being $\sigma \otimes \tau$, the middle identification being by the switch of the middle tensor factors, and the end map being $i_{U \otimes V} \otimes \mu$, where μ is the multiplication $A \otimes A \to A$. The composite $U \otimes V \to (U \otimes V) \otimes A$ is readily seen to be a comodule structure, and we call it the *tensor product comodule structure* of σ and τ. This corresponds to the usual tensor product of G-modules, i.e., it coincides with $(\sigma_* \otimes \tau_*)^*$.

Let us review the structures we have on $A = \mathscr{R}_F(G)$. First, A has the structure of an F-algebra with identity, and we denote the multiplication $A \otimes A \to A$ by μ. Second, A has the structure of an F-co-algebra, with comultiplication γ: $A \to A \otimes A$, and the F-algebra homomorphism c: $A \to F$ as a co-unit. Let s denote the linear automorphism of $A \otimes A$ such that $s(a_1 \otimes a_2) = a_2 \otimes a_1$. We regard $A \otimes A$ as an F-algebra, with $(\mu \otimes \mu) \circ (i_A \otimes s \otimes i_A)$ as the multiplication. We know that γ is then an F-algebra homomorphism.

Generally, if (C, γ, c) is an F-co-algebra, and (A, μ, u) is an F-algebra, where u is the map $F \to A$ obtained from the identity element of A, then we have the structure of an F-algebra on $\mathrm{Hom}_F(C, A)$, the multiplication (called *convolution*) being

$$(f, g) \to \mu \circ (f \otimes g) \circ \gamma$$

and having $u \circ c$ as its neutral element. In particular, if $C = A = \mathscr{R}_F(G)$, we have a convolution F-algebra structure (other than composition) on $\mathrm{Hom}_F(A, A)$, and the identity map i_A is invertible, i.e., there is an element η of $\mathrm{Hom}_F(A, A)$, called the *antipode* of A, such that

$$\mu \circ (\eta \otimes i_A) \circ \gamma = u \circ c = \mu \circ (i_A \otimes \eta) \circ \gamma .$$

In fact, η is given by $\eta(f)(x) = f(x^{-1})$ for every f in $\mathscr{R}_F(G)$ and every x in G.

The structure of A we have described so far, (A, μ, u, γ, c), is called the structure of a *Hopf algebra*. In the absence of a statement to the contrary, it will be understood here that the Hopf algebras considered possess an antipode. It can be shown that, in every Hopf algebra with antipode η, one has

$$\eta \circ \mu = \mu \circ (\eta \otimes \eta) \circ s ,$$

which means that η is necessarily an antimorphism of F-algebras $A \to A$. Similarly, η is necessarily an antimorphism of co-algebras, i.e.,

$$\gamma \circ \eta = s \circ (\eta \otimes \eta) \circ \gamma .$$

Finally, if the Hopf algebra is either commutative (as in our case) or co-commutative, then η is necessarily an involution, i.e., one has $\eta \circ \eta = i_A$. In our case, all these properties of η are easily verified by direct computation.

Let A be any Hopf algebra (not necessarily having an antipode) over F. From the comultiplication γ of A and the multiplication of F, we have the convolution F-algebra structure on $\mathrm{Hom}_F(A, F) = A^\circ$, as defined above. However, in the present case, the multiplication $F \otimes F \to F$ may be treated simply as an identification, and the product in A° of two elements x and y of A° will accordingly be written $(x \otimes y) \circ \gamma$. Clearly, this composition is bilinear, so that it gives A° the structure of an F-algebra. The identity element of this algebra is the co-unit c. This F-algebra A° is isomorphic with a certain subalgebra of the usual (composition) F-algebra $\mathscr{E}(A) = \mathrm{Hom}_F(A, A)$, as follows. Let us call a linear endomorphism e of A *proper* if

$$\gamma \circ e = (i_A \otimes e) \circ \gamma .$$

One verifies directly that the proper endomorphisms constitute an F-subalgebra of $\mathscr{E}(A)$, and that *the map $e \to c \circ e$ is an F-algebra isomor- of the algebra of all proper linear endomorphisms of A onto the F-algebra A°*. The inverse of this isomorphism sends every element σ of A° onto $(i_A \otimes \sigma) \circ \gamma$. Now let us suppose that our Hopf algebra has an antipode

η. Then one sees from a direct computation that every F-algebra homomorphism $h : A \to F$ is invertible in A°, the inverse being $h \circ \eta$. From the fact that γ is an F-algebra homomorphism, it follows that if x and y are F-algebra homomorphisms $A \to F$, so is their product $(x \otimes y) \circ \gamma$. Thus the F-algebra homomorphisms $A \to F$ constitute a subgroup of the group of units of A°. We denote this group by $\mathscr{G}(A)$. It can be verified directly that *the above F-algebra isomorphism of* A° *onto the algebra of proper linear endomorphisms of A maps* $\mathscr{G}(A)$ *onto the multiplicative group of all proper F-algebra automorphisms of A.*

In the case where $A = \mathscr{R}_F(G)$, the map that associates with every element x of G the evaluation x° of $\mathscr{R}_F(G)$ at x is a group homomorphism $G \to \mathscr{G}(\mathscr{R}_F(G))$. The proper F-algebra automorphism of A that corresponds to x is the left translation $f \to x \cdot f$ effected by x on $\mathscr{R}_F(G)$, and a linear endomorphism e of $\mathscr{R}_F(G)$ is proper if and only if $e(f \cdot x) = e(f) \cdot x$ for every f in $\mathscr{R}_F(G)$ and every x in G.

A subset T of $\mathscr{R}_F(G)$ is said to be *left stable* if $x \cdot T \subset T$ for every x in G. It is said to be *right stable* if $T \cdot x \subset T$ for every x in G. Finally, T is called *bistable* if it is both left and right stable, and T is called *fully stable* if it is bistable, and stable under the antipode η. An F-subalgebra of $\mathscr{R}_F(G)$ is a Hopf subalgebra if and only if it is fully stable. The stability conditions may be put in terms of the comultiplication as follows. An F-subspace S of $\mathscr{R}_F(G)$ is left stable if and only if $\gamma(S) \subset S \otimes \mathscr{R}_F(G)$, and it is right stable if and only if $\gamma(S) \subset \mathscr{R}_F(G) \otimes S$.

Any one of the usual representation theories, over the field F, of a given group G has its foundation in the choice of a particular Hopf subalgebra B of $\mathscr{R}_F(G)$, and it may then be viewed as the theory of B-comodules. Moreover, we may regard B as a Hopf algebra of representative functions on $\mathscr{G}(B)$ in the evident way. A group $\mathscr{G}(B)$, equipped with the Hopf algebra B of representative functions, is called a *pro-affine algebraic group.* If B is finitely generated as an F-algebra, then $\mathscr{G}(B)$ is an *affine algebraic group.* A pro-affine algebraic group may be viewed as a projective limit of affine algebraic groups (using the fact that B coincides with the union of the family of its finitely generated Hopf subalgebras), and many of the results for affine algebraic groups extend to pro-affine algebraic groups.

EXERCISES

1. Assuming the properties of the antipode as given in the text, show that every commutative Hopf algebra over a field coincides with the union of the family of its finitely generated Hopf subalgebras.

2. Let G be a finite group, F an arbitrary field. Make the Hopf algebra structure of $\mathscr{R}_F(G)$ explicit, using the following system of F-algebra generators: one function f_x for every element x of G, where $f_x(y)$ is equal to 1 or 0, according to whether $y = x$ or $y \neq x$.
3. Let $F[x]$ be the F-algebra of polynomials in one variable x with coefficients in the field F. Define an F-algebra homomorphism $\gamma : F[x] \to F[x] \otimes F[x]$ so that

$$\gamma(x) = x \otimes 1 + 1 \otimes x$$

and the F-algebra involution $\eta : F[x] \to F[x]$ so that $\eta(x) = -x$. Verify that these definitions make $F[x]$ into a Hopf algebra, and determine the group $\mathscr{G}(F[x])$.
4. Discuss the multiplicative analogue of Exercise 3, where $F[x]$ is replaced with $F[x, x^{-1}]$, and the formulas for γ and η are replaced with $\gamma(x) = x \otimes x$ and $\eta(x) = x^{-1}$.

3. AFFINE ALGEBRAIC GROUPS

The structure of an affine algebraic group over the field F consists of a pair (G, A), where G is a group, and A is a finitely generated fully stable F-subalgebra of the algebra $\mathscr{R}_F(G)$ of all F-valued representative functions on G, satisfying the following conditions: (1) A separates the elements of G; (2) every F-algebra homomorphism $A \to F$ is the evaluation $f \to x^\circ(f) = f(x)$ at an element x of G. These conditions mean that the canonical map $x \to x^\circ$ of G into $\mathscr{G}(A)$ is bijective, and we shall accordingly identify G with $\mathscr{G}(A)$. Clearly, any finitely generated Hopf algebra A over F whose elements are separated by $\mathscr{G}(A)$ defines the structure of an affine algebraic group $(\mathscr{G}(A), A)$.

Let H be a subgroup of the affine algebraic group $\mathscr{G}(A)$, and let I denote the annihilator of H in A, i.e., the ideal of the F-algebra A consisting of all elements f such that $x(f) = 0$ for every element x of H. We call H an *algebraic subgroup* of $\mathscr{G}(A)$ if the annihilator of I in $\mathscr{G}(A)$ coincides with H. It is easy to see that A/I inherits the structure of a Hopf algebra from A, and that if H is an *algebraic* subgroup of $\mathscr{G}(A)$, then H may be identified with the affine algebraic group $\mathscr{G}(A/I)$ having A/I as its Hopf algebra of functions. Actually, we shall record a more precise result concerning general Hopf algebras. If A is any Hopf algebra (not necessarily being commutative, and not necessarily possessing an antipode) with comultiplication γ and co-unit c, then a *Hopf ideal* of A is a two-sided ideal J with the additional properties that

$$\gamma(J) \subset J \otimes A + A \otimes J,$$

and $c(J) = (0)$.

Proposition 3.1. *Let A be a Hopf algebra over the field F, with comultiplication γ and co-unit c. Let G denote the monoid of all F-algebra homomorphisms $A \to F$. If J is a Hopf ideal of A then the annihilator of J in G is a submonoid H of G. An element x of G belongs to H if and only if*

$(i_A \otimes x) \circ \gamma$ stabilizes J, or equivalently if and only if $(x \otimes i_A) \circ \gamma$ stabilizes J. If A has an antipode η, so that G is a group, then H is a subgroup of G, and an element x of G belongs to H if and only if $(i_A \otimes x)(\gamma(J)) = J$, or equivalently if and only if $(x \otimes i_A)(\gamma(J)) = J$.

Conversely, if H is a submonoid of G, then the annihilator I of H in A is a Hopf ideal. Moreover, if A has an antipode η, then $\eta(I) \subset I$.

Proof: Let J be a Hopf ideal of A, and let H be its annihilator in G. Evidently, c belongs to H. From $\gamma(J) \subset J \otimes A + A \otimes J$, one sees immediately that $(i_A \otimes x) \circ \gamma$ and $(x \otimes i_A) \circ \gamma$ stabilize J whenever x belongs to H. Conversely, suppose that x is an element of G such that $(i_A \otimes x) \circ \gamma$ stabilizes J. Then $c \circ (i_A \otimes x) \circ \gamma$ annihilates J. But

$$c \circ (i_A \otimes x) \circ \gamma = x \circ (c \otimes i_A) \circ \gamma = x\,,$$

so that x belongs to H. Similarly, we see that if $(x \otimes i_A) \circ \gamma$ stabilizes J then x belongs to H.

If x and y are elements of H, then $(i_A \otimes x) \circ \gamma$ and $(i_A \otimes y) \circ \gamma$ stabilize J, so that their composite also stabilizes J. But this composite is equal to $(i_A \otimes (x \otimes y) \circ \gamma) \circ \gamma$. By what we have already proved, it follows that $(x \otimes y) \circ \gamma$ belongs to H. Thus H is a submonoid of G.

Now suppose that A has an antipode η. Then, for each x in G, the map $(i_A \otimes x) \circ \gamma$ is an algebra automorphism of A, its inverse being $(i_A \otimes x \circ \eta) \circ \gamma$, and the map $x \to (i_A \otimes x) \circ \gamma$ is a representation of the group G on A, with respect to which A is locally finite. In particular, if J and H are as above, and if x is an element of H, then $(i_A \otimes x) \circ \gamma$ acts as an injective linear endomorphism on J, and J is the sum of finite-dimensional $(i_A \otimes x) \circ \gamma$-stable subspaces. It follows that $(i_A \otimes x) \circ \gamma$ is also surjective on J, and that its inverse $(i_A \otimes x \circ \eta) \circ \gamma$ stabilizes J. Thus $(i_A \otimes x)(\gamma(J)) = J$, and the inverse $x \circ \eta$ of x in G belongs to H, so that H is a subgroup of G. The fact that, for x in H, we have $(x \otimes i_A)(\gamma(J)) = J$ is proved in exactly the same manner.

Now let H be any submonoid of G, and let I be the annihilator of H in A. Evidently, I is a two-sided ideal of A, and $c(I) = (0)$. Now let (a_q) be a set of elements of A whose canonical images in A/I constitute an F-basis of A/I. Then, if b is any element of I, we may write

$$\gamma(b) = \sum_q u_q \otimes a_q + k\,,$$

where k belongs to $A \otimes J$, and each u_q is an element of A. Let x and y be elements of H. Since H is a submonoid of G, we have then

$$(x \otimes y) \circ \gamma \in H\,,$$

so that $(x \otimes y)(\gamma(b)) = 0$, i.e., $\sum_q x(u_q)y(a_q) = 0$. Keeping x fixed, and letting y range over H, we conclude from this that $\sum_q x(u_q)a_q$ belongs to I. By the choice of the a_q's, this implies that $x(u_q) = 0$ for every index q. Since this holds for every x in H, we conclude that each u_q belongs to I. Thus $\gamma(b)$ belongs to $I \otimes A + A \otimes I$, and we have shown that I is a Hopf ideal of A.

Finally, suppose that A has an antipode η. Then we know that the annihilator of I in G is a subgroup of G. If x is an element of H, its inverse $x \circ \eta$ therefore annihilates I, which shows that $\eta(I) \subset I$. This completes the proof of Proposition 3.1.

Before we proceed further, we introduce some terminology. If $G = \mathcal{G}(A)$ is an affine algebraic group, then we call the associated Hopf algebra A the algebra of *polynomial functions* on G, and we sometimes write $\mathcal{A}(G)$ for A. For instance, if H is an algebraic subgroup of G, and if I is the annihilator of H in A, then $\mathcal{A}(H)$ is the Hopf algebra A/I. A *morphism of affine algebraic groups* is a group homomorphism $\rho : G \to K$ such that $\mathcal{A}(K) \circ \rho \subset \mathcal{A}(G)$. The map $f \to f \circ \rho$ of $\mathcal{A}(K)$ into $\mathcal{A}(G)$ is then a morphism of Hopf algebras, i.e., an F-algebra homomorphism that is at the same time an F-co-algebra homomorphism, compatible with the co-units, in the evident sense. Conversely, a morphism of Hopf algebras $\mathcal{A}(K) \to \mathcal{A}(G)$ induces a morphism of affine algebraic groups $G \to K$ in the natural fashion. The injection map $H \to G$ of an algebraic subgroup H of G is evidently a morphism of affine algebraic groups. The corresponding morphism of Hopf algebras is the canonical map $A \to A/I$, where I is the annihilator of H in $A = \mathcal{A}(G)$.

The main significance of Proposition 3.1 resides in the following application. Let G be an affine algebraic group, and let A be its algebra of polynomial functions. Let H be a submonoid of G, and let I be the annihilator of H in A. Let $[H]$ denote the annihilator of I in G. Then I is evidently the annihilator of $[H]$ in A, and we know from Proposition 3.1 that $[H]$ is a subgroup of G. Clearly, $[H]$ is an *algebraic* subgroup of G. We call $[H]$ the *algebraic hull* of H in G, and we observe that it coincides with the algebraic hull of the subgroup of G that is generated by H.

Let G and H be affine algebraic groups over F, and let A and B be their algebras of polynomial functions. The Hopf algebra structures of A and B yield a Hopf algebra structure of $A \otimes B$ in the natural fashion, and we may consider the group $\mathcal{G}(A \otimes B)$ of all F-algebra homomorphisms $A \otimes B \to F$. The elements of $\mathcal{G}(A \otimes B)$ are evidently the F-algebra homomorphisms $x \otimes y$, with x ranging over $\mathcal{G}(A)$, and y over

$\mathscr{G}(B)$. One verifies directly that the map $(x, y) \to x \otimes y$ is a group isomorphism of $\mathscr{G}(A) \times \mathscr{G}(B)$, i.e., of $G \times H$, onto $\mathscr{G}(A \otimes B)$, and that $\mathscr{G}(A \otimes B)$ separates the elements of $A \otimes B$. Identifying $G \times H$ with $\mathscr{G}(A \otimes B)$, we have therefore the structure of an affine algebraic group on $G \times H$, whose algebra of polynomial functions is $\mathscr{A}(G) \otimes \mathscr{A}(H)$. The canonical injections $G \to G \times H$ and $H \to G \times H$, as well as the canonical projections $G \times H \to G$ and $G \times H \to H$, are clearly morphisms of affine algebraic groups. A pair $K \to G$ and $K \to H$ of morphisms of affine algebraic groups evidently yields one and only one morphism of affine algebraic groups $K \to G \times H$ whose composites with the projections are the given morphisms. Thus the above is the appropriate definition of direct product in the category of affine algebraic groups.

Let L be an extension field of our base field F, and let (G, A) be the structure of an affine algebraic group over F. By linear extension, the Hopf algebra structure of A yields the structure of an L-Hopf algebra on $A \otimes L$. The elements of $G = \mathscr{G}(A)$ yield L-algebra homomorphisms $A \otimes L \to L$ in the natural way, so that we have a canonical injective group homomorphism $\mathscr{G}(A) \to \mathscr{G}(A \otimes L)$. It is easy to see that the image of $\mathscr{G}(A)$ in $\mathscr{G}(A \otimes L)$ separates the elements of $A \otimes L$. Hence $\mathscr{G}(A \otimes L)$ is an affiine algebraic group over L, with $A \otimes L$ as its algebra of polynomial functions, and the image of G in $\mathscr{G}(A \otimes L)$ is *algebraically dense* in $\mathscr{G}(A \otimes L)$, i.e., its algebraic hull in $\mathscr{G}(A \otimes L)$ coincides with $\mathscr{G}(A \otimes L)$. Usually, one writes G^L for $\mathscr{G}\big(\mathscr{A}(G) \otimes L\big)$, and one calls it the *extension of G over L*. It is readily verified that, if H is an algebraic subgroup of G, then H^L may be identified with the algebraic hull of the canonical image of H in G^L, and the annihilator of H^L in $A \otimes L$ is $I \otimes L$, where I is the annihilator of H in A. If $\rho: G \to K$ is a morphism of affine algebraic groups over F, then the corresponding morphism of Hopf algebras $\mathscr{A}(K) \to \mathscr{A}(G)$ extends naturally to a morphism of Hopf algebras $\mathscr{A}(K) \otimes L \to \mathscr{A}(G) \otimes L$. In turn, this defines a morphism $\rho^L: G^L \to K^L$ of affine algebraic groups over L, which is called the extension of ρ over L.

EXERCISES

1. Let E be a finite-dimensional associative algebra with identity element over a field F. Let $U(E)$ denote the multiplicative group of units of E. Let A be the smallest Hopf algebra of representative F-valued functions on $U(E)$ containing the restrictions to $U(E)$ of the linear

functions $E \to F$. Show that $(U(E), A)$ is the structure of an affine algebraic group.

2. In the notation of Exercise 1, let ρ be a faithful representation of E by linear endomorphisms of a finite-dimensional F-space V. For an element e of $U(E)$, let $d(e)$ be the determinant of the linear automorphism $\rho(e)$. Show that the algebra A of Exercise 1 is generated by the representative functions associated with the restriction of ρ to $U(E)$ (in the sense of Section 2) and the reciprocal of the function d. By choosing a basis of V and considering the corresponding matrix entry functions, establish a connection between the comultiplication of A and the usual rule for multiplying matrices.
3. Let (G, A) be the structure of an affine algebraic group. Prove that the center of G is an algebraic subgroup of G.

4. DECOMPOSITION INTO COMPONENTS

Let (G, A) be the structure of an affine algebraic group over the field F. We define an ideal P of A as follows. An element a of A is to belong to P if and only if there is an element a' in A such that $aa' = 0$ and a' is not annihilated by the co-unit c of A. Using the fact that c is an F-algebra homomorphism $A \to F$, one sees readily that P is actually an ideal of A. Now A is a Noetherian ring, so that there is a finite set $(p_1, \ldots, p_n)$ of elements of P such that $P = Ap_1 + \cdots + Ap_n$. For each p_i, choose an element p_i' in A as in the above definition of P. Put $q = p_1' \cdots p_n'$. Then we have $Pq = (0)$ and $c(q) \neq 0$, so that P is precisely the annihilator of q in A.

Now let G_1 denote the annihilator of P in G. We wish to show that G_1 is a subgroup of G. Evidently, the neutral element c of G belongs to G_1. Moreover, if x belongs to G_1, so does its inverse $x \circ \eta$, because the antipode η evidently stabilizes P. Now let p be an element of P, and let y be an element of G. To the relation $pq = 0$, apply the algebra automorphism $(y \otimes i_A) \circ \gamma$. In order to simplify the notation, let us view A as an algebra of representative functions on G, and let us note that $(y \otimes i_A) \circ \gamma$ is the right translation effected by y on A. Thus the above gives $(p \cdot y)(q \cdot y) = 0$. If $c(q \cdot y) \neq 0$, i. e., if $q(y) \neq 0$, this shows that $p \cdot y$ belongs to P. Hence, in any case, we have $(p \cdot y)(x)q(y) = 0$ for every y in G and every x in G_1. This may be written in the form $\big((x \cdot p)q\big)(y) = 0$, so that we obtain $(x \cdot p)q = 0$ for every x in G_1. Thus we have shown that $x \cdot P \subset P$ for every x in G_1. Evidently, this implies that G_1 is a submonoid of G. Since G_1 is also stable under the inversion of G, it is therefore a subgroup of G.

Next, we show that the annihilator of G_1 in A coincides with P. In order to do this, it suffices to show that if u is an element of A that does not belong to P, then there is an element x in G_1 such that $x(u) \neq 0$. Now we have $uq \neq 0$, so that there is an element x in G such that $x(uq) \neq 0$, i.e., $x(u) \neq 0$ and $x(q) \neq 0$. Since $Pq = (0)$ and $x(q) \neq 0$, this element x must belong to G_1. Hence the annihilator of G_1 in A indeed coincides with P.

Taken together, our results so far mean that G_1 is an algebraic subgroup of G, and that the algebra of polynomial functions on G_1 is A/P.

An affine algebraic group is said to be *connected* if its algebra of polynomial functions is an integral domain. The full result concerning G_1 that we wish to establish is as follows.

THEOREM 4.1. *The algebraic subgroup G_1 of the affine algebraic group G, as defined above, is normal and of finite index in G. As an affine algebraic group, G_1 is connected, and G_1 is the only connected algebraic subgroup of finite index in G.*

Proof: Let x be an element of G, and consider the conjugation $a \to x \cdot a \cdot x^{-1}$ effected by x on the function algebra A. This is an F-algebra automorphism of A, and it evidently stabilizes the kernel of the co-unit c. Hence it is clear from the definition of the annihilator P of G_1 that this conjugation also stabilizes P. Since G_1 is the annihilator of P in G, this implies that $x^{-1}G_1x \subset G_1$, whence G_1 is normal in G.

Next, we show that G_1 is connected, i. e., that P is a prime ideal. Let u and v be elements of A such that uv belongs to P, but v does not belong to P. Choose an element x in G_1 such that $x(v) \neq 0$. Now $(x \cdot u)(x \cdot v)$ belongs to P, so that $(x \cdot u)(x \cdot v)q = 0$, where q is the element used above in our initial discussion of P. Since $c\big((x \cdot v)q\big) = x(v)c(q) \neq 0$, this shows that $x \cdot u$ belongs to P, whence also u belongs to P, as had to be shown.

In order to prove that G_1 is of finite index in G, we use the fact that q is a representative function on G to choose a finite set $(x_1, \ldots, x_n)$ of elements of G such that every translate $x \cdot q$ is an F-linear combination of the n translates $x_i \cdot q$. Since $x^{-1}(x \cdot q) = c(q) \neq 0$, there in an index i such that $x^{-1}(x_i \cdot q) \neq 0$. This means that the element $x^{-1}x_i$ of G does not annihilate q. Since P is the annihilator of q, this implies that $x^{-1}x_i$ belongs to G_1. Thus we have $G = \bigcup_{i=1}^{n} x_iG_1$, showing that G/G_1 is finite.

Finally, let H be any connected algebraic subgroup of finite index in G. Let I denote the annihilator of H in A. Then I is a prime ideal, and q does not belong to I because $c(q) \neq 0$. It follows that we must have

$P \subset I$. Let $H, y_1H, \ldots, y_mH$ be all the cosets of H of the form xH. Then none of the elements y_i^{-1} belongs to H, so that there is an element f_i in I such that $y_i^{-1}(f_i) \neq 0$. Let f be the product of the m translates $f_i \cdot y_i^{-1}$. Then f vanishes on every coset y_iH, while $c(f) \neq 0$. Let g be any element of I. Because of the vanishing property of f, we have $fg = 0$. Since $c(f) \neq 0$, this implies that g belongs to P. Thus we have $I \subset P$, so that (from the above) $I = P$. Hence $H = G_1$, and the proof of Theorem 4.1 is complete.

The group G_1 is called the *connected component of the neutral element in* G.

PROPOSITION 4.2. *Let G and H be connected affine algebraic groups over the field F, and let L be a field containing F. Then the extension G^L of G over L, and the direct product $G \times H$, are connected affine algebraic groups.*

Proof: Let A and B denote the algebras of polynomial functions on G and H, respectively. Then the algebra of polynomial functions on G^L is $A \otimes L$, and the algebra of polynomial functions on $G \times H$ is $A \otimes B$. Hence it suffices to show that, if C is any F-algebra that is an integral domain, then $A \otimes C$ is an integral domain. In order to do this, let u and v be elements $A \otimes C$ such that $uv = 0$. Choose a finite F-linearly independent set $(c_1, \ldots, c_n)$ of elements of C such that

$$u = \sum_{i=1}^{n} u_i \otimes c_i , \qquad \text{and} \qquad v = \sum_{i=1}^{n} v_i \otimes c_i ,$$

where the u_i's and the v_i's belong to A. Let x be an element of G, and regard x also as an F-algebra homomorphism $A \otimes C \to C$, via C-linear extension. If there is an index j such that $x(v_j) \neq 0$ then we have $x(v) \neq 0$, and therefore $x(u) = 0$, because $x(u)x(v) = 0$. This gives $x(u_i) = 0$ for each i. Hence, in any case, we have $x(u_i)x(v_j) = 0$ for every i, every j, and every x in G. Therefore, each $u_iv_j = 0$. If $v \neq 0$, then one of the v_j's must be different from 0, so that then each u_i must be 0, whence $u = 0$. This proves Proposition 4.2.

PROPOSITION 4.3. *Let G be an affine algebraic group over the field F, and let L be a field containing F. Let G_1 denote the connected component of the neutral element in G, and let $(x_1, \ldots, x_n)$ be a complete system of representatives in G for the cosets of G_1 in G. Then $G_1{}^L$ is the connected component of the neutral element in G^L, and $(x_1, \ldots, x_n)$ is a complete system of representatives in G^L for the cosets of $G_1{}^L$ in G^L.*

Proof: As is already implied by the statement of the proposition, we identify G and G_1 with their canonical images in G^L, and we note that $G_1{}^L$, when regarded as a subgroup of G^L, is then the algebraic hull of G_1 in G^L. By Proposition 4.2, $G_1{}^L$ is connected. We show that the cosets $x_iG_1{}^L$ are mutually disjoint, as follows. Suppose that $x_iG_1{}^L$ and $x_jG_1{}^L$ have an element in common, i. e., that there are elements y_i and y_j in $G_1{}^L$ such that $x_iy_i = x_jy_j$. Then we have $x_i{}^{-1}x_j \in G_1{}^L \cap G = G_1$. Hence we must have $i = j$, as had to be shown.

Now let y be any element of G^L, and let us show that y belongs to one of the cosets $x_iG_1{}^L$. Suppose this is not the case. Then none of the elements $x_i^{-1}y$ belongs to $G_1{}^L$. Hence, for each i, there is an element a_i of the annihilator of G_1 in the algebra A of polynomial functions on G such that $(x_i^{-1}y)(a_i) \neq 0$, i. e., $y(a_i \cdot x_i^{-1}) \neq 0$. On the other hand, since G is the union of the family of the x_iG_1's, the product in A of the translates $a_i \cdot x_i^{-1}$ is annihilated by every element of G, and is therefore 0, which contradicts the above. Thus we may conclude that G^L is the union of the $x_iG_1{}^L$'s so that $G_1{}^L$ is a connected algebraic subgroup of finite index in G^L. By Theorem 4.1, this implies that $G_1{}^L$ is the connected component of the neutral element in G^L, which completes the proof of Proposition 4.3.

The decomposition of G into the cosets of G_1 leads to a direct sum decomposition of the F-algebra A of polynomial functions. Choose $(x_1, \ldots, x_n)$ as in Proposition 4.3, making x_1 the neutral element of G. Let P denote the annihilator of G_1 in A, and let L be an algebraically closed extension field of F. The annihilator of $G_1{}^L$ in $A \otimes L$ is $P \otimes L$. Hence the annihilator of $x_iG_1{}^L$ in $A \otimes L$ is $(P \otimes L) \cdot x_i^{-1} = P \cdot x_i^{-1} \otimes L$. Put $Q = \bigcap_{i \neq 1} P \cdot x_i^{-1}$. Then $Q \otimes L = \bigcap_{i \neq 1} (P \otimes L) \cdot x_i^{-1}$, and the annihilator of $Q \otimes L$ in G^L is $\bigcup_{i \neq 1} x_iG_1{}^L$. Since G^L is the disjoint union of the family of cosets $x_iG_1{}^L$ (by Proposition 4.3), it follows that the ideal $P \otimes L + Q \otimes L$ of $A \otimes L$ has no zero in G^L. Since L is algebraically closed, this implies that $P \otimes L + Q \otimes L = A \otimes L$; otherwise this ideal would be contained in some maximal ideal M, and the canonical L-algebra homomorphism $A \otimes L \to (A \otimes L)/M$ could be followed by an L-algebra homomorphism $(A \otimes L)/M \to L$ (which exists by virtue of Theorem 1.3) to yield an element of G^L annihilating $P \otimes L + Q \otimes L$. It follows at once from this that we must already have $P + Q = A$. In particular, $1 = e_1 + f_1$, with e_1 in P and f_1 in Q. As an F-valued function on G, f_1 vanishes on each coset x_iG_1 other than G_1, while f_1 reduces to the constant function with value 1 on G_1. For each i, let $f_i = f_1 \cdot x_i^{-1}$.

Then $(f_1, \ldots, f_n)$ is a system of mutually orthogonal nonzero idempotents in A, and A is the direct sum of the F-algebras Af_i. The summand Af_1 may be identified with the algebra of all polynomial functions of G_1. Summarizing and simplifying, we may state our result as follows, where we take into account the normality of G_1 in G.

THEOREM 4.4 *Let (G, A) be the structure of an affine algebraic group, and let G_1 denote the connected component of the neutral element in G. Then A is a direct algebra sum $A_1 + \cdots + A_n$, such that A_1 may be identified with $\mathscr{A}(G_1)$ by the restriction map, each A_i is stable under the translation actions of G_1, every translation effected by an element of G permutes the A_i's and G acts transitively on the set of the A_i's via left translations, as well as via right translations.*

EXERCISES

1. Let G be an affine algebraic group, and let G_1 be the connected component of the neutral element in G. Show that every subgroup of G that contains an algebraic subgroup of G as a subgroup of finite index is an algebraic subgroup of G. Hence prove that a subgroup H of G is an algebraic subgroup of G if and only if $H \cap G_1$ is an algebraic subgroup of G_1.
2. Let (G, A) be the structure of an affine algebraic group over the field F. Let G_1 be the connected component of the neutral element in G, and let $A = A_1 + \cdots + A_n$ be the decomposition of Theorem 4.4. Write $1 = f_1 + \cdots + f_n$, with each f_i in A_i. Show that the sum $Ff_1 + \cdots + Ff_n$ is a Hopf subalgebra of A, and that it may be identified with the Hopf algebra of all representative functions on the group G/G_1.

5. POLYNOMIAL MAPS AND ALGEBRAIC SUBGROUPS

Let G and H be affine algebraic groups. A map $\rho: G \to H$ is called a *polynomial map* if $\mathscr{A}(H) \circ \rho \subset \mathscr{A}(G)$. The main purpose of this section is to prove the following theorem.

THEOREM 5.1 *Let G and H be affine algebraic groups over an algebraically closed field F. Let $\rho: G \to H$ be a polynomial map that sends the neutral element of G onto that of H, and assume that G is connected. Then the products of finite sequences of elements of $\rho(G)$ constitute a connected algebraic subgroup T of H, and there is a natural number n such that every element of T is the product of n elements of $\rho(G)$.*

Proof: For every positive natural number m, let G_m denote the direct product of m copies of G. Let ρ_m be the map of G_m into H defined by $\rho_m(x_1, \ldots, x_m) = \rho(x_1) \cdots \rho(x_m)$. Clearly, ρ_m is a polynomial map. Let J_m denote the annihilator of $\rho_m(G_m)$ in $\mathscr{A}(H)$. Then J_m is the kernel of the F-algebra homomorphism of $\mathscr{A}(H)$ into $\mathscr{A}(G_m)$ that sends every element g of $\mathscr{A}(H)$ onto $g \circ \rho_m$. Since G is connected, we know from Proposition 4.2 that G_m is connected. Hence J_m is a prime ideal of $\mathscr{A}(H)$. Since $\rho_m(G_m) \subset \rho_{m+1}(G_{m+1})$, we have $J_{m+1} \subset J_m$. By Theorem 4.4, $\mathscr{A}(H)$ is the direct F-algebra sum of a finite family of finitely generated F-algebras that are integral domains. Hence it follows from Proposition 1.8 that there is an upper bound for the lengths of prime ideal chains in $\mathscr{A}(H)$. In particular, there must be a natural number q such that $J_m = J_q$ for every $m \geqq q$. Now J_q is the annihilator in $\mathscr{A}(H)$ of the submonoid of H consisting of all products of elements of $\rho(G)$. Let T be the annihilator of J_q in H. From Proposition 3.1, we have that T is a subgroup of H, so that T is an algebraic subgroup of

H, its annihilator in $\mathscr{A}(H)$ being J_q. Since J_q is a prime ideal, T is connected.

Noting that F is algebraically closed and applying Theorem 1.3, we see that there is a non-zero element f in $\mathscr{A}(H) \circ \rho_q$ such that every F-algebra homomorphism $\mathscr{A}(H) \circ \rho_q \to F$ not annihilating f extends to an F-algebra homomorphism $\mathscr{A}(G_q) \to F$, i.e., is the restriction of an element of G_q. Write $f = g \circ \rho_q$, with g in $\mathscr{A}(H)$. Since $f \neq 0$, this element g does not belong to J_q. Now let x be any element of T. Since J_q is stable under the antipode η, as well as under translations with elements of T, the element $x \cdot \eta(g)$ does not belong to J_q. Hence $(x \cdot \eta(g)) \circ \rho_q \neq 0$, so that $\rho_q(u)(x \cdot \eta(g)) \neq 0$ for some element u in G_q, which means that $g(x^{-1}\rho_q(u)^{-1}) \neq 0$. Since $x^{-1}\rho_q(u)^{-1}$ belongs to T, it annihilates J_q, and therefore induces an F-algebra homomorphism $\sigma: \mathscr{A}(H) \circ \rho_q \to F$ where $\sigma(h \circ \rho_q) = h(x^{-1}\rho_q(u)^{-1})$ for every element h of $\mathscr{A}(H)$. In particular, $\sigma(f) = g(x^{-1}\rho_q(u)^{-1}) \neq 0$. By the choice of f, the homomorphism σ is therefore the restriction of an element v of G_q. Hence, for every h in $\mathscr{A}(H)$, we have $h(x^{-1}\rho_q(u)^{-1}) = h(\rho_q(v))$, so that $x^{-1}\rho_q(u)^{-1} = \rho_q(v)$, or $x^{-1} = \rho_q(v)\rho_q(u)$. This shows that every element of T is the product of $2q$ elements of $\rho(G)$, and Theorem 5.1 is proved.

The following three corollaries are immediate consequences.

COROLLARY 5.2 *Let G and H be affine algebraic groups over an algebraically closed field, with G connected. Let $\rho: G \to H$ be a morphism of affine algebraic groups. Then $\rho(G)$ is a connected algebraic subgroup of H.*

COROLLARY 5.3 *Let G be a connected affine algebraic group over an algebraically closed field. Then the commutator subgroup $[G, G]$ of G is a connected algebraic subgroup of G, and there is a natural number m such that every element of $[G, G]$ is a product of m commutators $xyx^{-1}y^{-1}$.*

COROLLARY 5.4 *Let U and V be connected algebraic subgroups of an affine algebraic group G over an algebraically closed field. Then the products of finite sequences of elements of $U \cup V$ constitute a connected algebraic subgroup of G, and there is a natural number m such that every element of this group is a product of m elements of $U \cup V$.*

EXERCISES

1. Extend Corollary 5.2 to the case where G is not necessarily connected, making use of Exercise 1 of Section 4.

2. For an arbitrary group G, note that every element of the commutator subgroup $[G, G]$ can be written in the form $x_1 \cdots x_k x_1^{-1} \cdots x_k^{-1}$, as follows inductively from the identity

$$x_1 \cdots x_m x_1^{-1} \cdots x_m^{-1} x y x^{-1} y^{-1}$$
$$= \left(x_1 \cdots x_m x^{-1} y^{-1}(yx)\right)\left(x_1^{-1} \cdots x_m^{-1} x y (yx)^{-1}\right).$$

Use this and Corollary 5.3 to show that if G is a connected affine algebraic group over an algebraically closed field then there is a natural number q such that every element of $[G, G]$ is of the form $x_1 \cdots x_q x_1^{-1} \cdots x_q^{-1}$.

6. FACTOR GROUPS

The following theorem is essential to the discussion of factor groups. Roughly, its content is that an algebraic subgroup of an affine algebraic group is defined by its semi-invariants.

Theorem 6.1. *Let (G, A) be the structure of an affine algebraic group over an arbitrary field F, and let H be an algebraic subgroup of G. There is a finite subset P of A, and an element f of A whose restriction to H is a group homomorphism of H into the multiplicative group of F, such that*

(1) *for every p in P and every x in H, $x \cdot p = f(x)p$;*

(2) *every element x of G such that $x \cdot p \in Fp$, for every element p of P, belongs to H.*

Proof: Let I denote the annihilator of H in A. We can find a finite-dimensional left stable F-subspace V of A such that $V \cap I$ generates I as an ideal. Let d denote the dimension of $V \cap I$, and let W be the homogeneous component of degree d of the exterior F-algebra built on V. The left translation action of G on V canonically yields a representation of G on W whose associated representative functions evidently belong to A. Let S denote the F-subspace of W that is spanned by the exterior products of sequences formed from d elements of $V \cap I$. Clearly, S is 1-dimensional. Choose a non-zero element s of S, so that $S = Fs$.

If σ is any linear functional on W, then we shall denote by σ/s the F-valued function on G that is given by

$$(\sigma/s)(x) = \sigma(x \cdot s)\,,$$

where $x \cdot s$ denotes the x-transform of s with respect to our representation of G on W, so that σ/s is an element of A.

Since $H \cdot I \subset I$, it is clear that S is H-stable. Hence, for every element x of H, we have $x \cdot s = g(x)s$, where g is a group homomorphism of H into the multiplicative group of F. Choose a linear functional σ on W

such that $\sigma(s) = 1$, and put $f = \sigma/s$. Then f belongs to A, and the restriction of f to H coincides with g.

Now let $(\sigma_1, \ldots, \sigma_p)$ be an F-basis for the annihilator of S in the dual of W, and consider the elements σ_i/s of A. If x belongs to H, we have

$$x \cdot (\sigma_i/s) = \sigma_i/(x \cdot s) = g(x)(\sigma_i/s) .$$

Conversely, suppose that x is an element of G such that $x \cdot (\sigma_i/s)$ is an F-multiple of σ_i/s for each i. Evaluating at the neutral element of G, we see that this implies that $\sigma_i(x \cdot s) = 0$ for each i, whence $x \cdot s$ belongs to S. Hence we have $x \cdot S = S$. Now let t be any element of $V \cap I$. Then the exterior product tS is (0), whence its x-transform $(x \cdot t)(x \cdot S)$ is (0), i.e., $(x \cdot t)S = (0)$. But this means that $x \cdot t$ belongs to $V \cap I$. Thus we have $x \cdot (V \cap I) \subset V \cap I$. Since $V \cap I$ generates I as an ideal, this implies that x belongs to H. Theorem 6.1 is therefore established, with P the set of elements σ_i/s.

If p is an element of A, and g is a homomorphism of H into the multiplicative group of F such that $x \cdot p = g(x)p$ for every element x of H, then p is called a *semi-invariant* of H, and g is called the *weight* of p.

If Q is any subset of A then the *left fixer* of Q in G is the set of all elements x of G such that $x \cdot q = q$ for every q in Q. Evidently, this is an algebraic subgroup of G. The *right fixer* of Q in G is defined analogously. It is readily seen that if the left fixer of Q in G is a *normal* subgroup of G then it is contained in the right fixer of Q in G, and hence is the *fixer* of Q in G, i.e., is the set of all elements x of G such that $x \cdot q = q = q \cdot x$ for every q in Q.

THEOREM 6.2. *Let (G, A) be the structure of an affine algebraic group over an arbitrary field F, and let H be a normal algebraic subgroup of G. Then there is a finite subset Q of A such that the left fixer of Q in G is precisely H.*

Proof: Let P be a finite set of semi-invariants of H, such as is given by Theorem 6.1, and let g denote the common weight of the elements of P. Let J be the smallest left G-stable subspace of A that contains P. The elements of J that are H-semi-invariants of weight g evidently constitute an H-submodule J_1 of J. If x is any element of G, then $x \cdot J_1$ is clearly the H-submodule of J consisting of all elements of J that are semi-invariants of weight g_x, where $g_x(y) = g(x^{-1}yx)$. Since J is finite-dimensional, it is a finite direct H-module sum $J_1 + \cdots + J_k$, where the J_i's are the distinct $x \cdot J_1$'s.

Let E denote the F-algebra of all F-linear endomorphisms of J. Let

E' denote the subalgebra of E consisting of the endomorphisms that stabilize each J_i. Let $\rho: G \to E$ be the representation of G by left translations on J. It is clear from the definitions that, if e is an element of E' and x is an element of G, then $\rho(x)e\rho(x)^{-1}$ is again an element of E'. Thus we have a representation σ of G on E', where

$$\sigma(x)(e) = \rho(x)e\rho(x)^{-1}.$$

It is readily verified that the representative functions associated with σ belong to A (in the notation of p. 15, $\mathscr{S}(\rho)$ is the smallest bistable subspace of A containing J, and $\mathscr{S}(\sigma) \subset \mathscr{S}(\rho)\eta(\mathscr{S}(\rho))$, where η is the antipode). Since every element of H acts as a scalar multiplication on each J_i, the kernel of σ contains H, so that the representative functions associated with σ must actually belong to the left H-fixed part A^H of A (since H is normal in G, this is also the right H-fixed part of A). Let Q be any finite subset of A^H that spans the space of representative functions associated with σ. Let x be any element of the left fixer of Q in G. Then x must evidently belong to the kernel of σ, which means that $\rho(x)$ commutes with every element of E'. It follows at once from this that $\rho(x)$ stabilizes each J_i, and that the restriction of $\rho(x)$ to J_i is a scalar multiplication. Since $P \subset J_1$, the element x therefore satisfies condition (2) of Theorem 6.1, whence x belongs to H. This proves Theorem 6.2.

If A is any commutative ring, we denote by $[A]$ the *total ring of fractions* of A, i.e., the ring of fractions formed with respect to the set of all non-zerodivisors of A. If A is an algebra of representative functions on a group G, then the translation actions of G on A evidently extend uniquely to actions of G by algebra automorphisms of $[A]$.

LEMMA 6.3. *Let (G, A) be the structure of an affine algebraic group over an arbitrary field F. Suppose that f is an element of $[A]$ such that the transforms $x \cdot f$, with x ranging over G, all lie in a finite-dimensional F-subspace of $[A]$. Then f belongs to A.*

Proof: Let $(f_1, \ldots, f_n)$ be an F-basis for the space spanned by the transforms $x \cdot f$. There is a non-zerodivisor t in A such that each tf_i belongs to A, and then we have $t(x \cdot f)$ in A for every element x of G. Equivalently, $(x \cdot t)f$ is in A for every x in G. Now let L be an algebraically closed extension field of F, identify A with its canonical image in $A \otimes L$, and identify G with its canonical image in $G^L = \mathscr{G}(A \otimes L)$. Since G is algebraically dense in G^L, there is, for every y in G^L, an element x in G such that $t(yx) \neq 0$, i.e., such that $(x \cdot t)(y) \neq 0$. Let J be the ideal of $A \otimes L$ consisting of all elements u such that uf lies in $A \otimes L$. Our last conclusion shows that J has no zero in G^L. As in our proof of Theorem 4.4, we conclude from this that $J = A \otimes L$, whence f belongs to $A \otimes L$. Using the fact that f belongs to $[A]$, we see from this that f actually belongs to A, proving Lemma 6.3.

THEOREM 6.4. *Let (G, A) be the structure of an affine algebraic group over an algebraically closed field F, and let B be any fully stable F-subalgebra of A. Then B is finitely generated as an F-algebra, and every F-algebra homomorphism $B \to F$ is the restriction to B of an element of G.*

Proof: Let H denote the left fixer of B in G. Let b be an element of B, and let x be an element of H. Then $x \cdot b = b$, i.e., $b(yx) = b(y)$ for every element y of G, which means that the element $b \cdot y - b(y)$ of B takes the value 0 at x, for every y in G. Hence we see that if I denotes the ideal of A that is generated by the elements $b \cdot y - b(y)$ with b in B, and y in G, then H is the annihilator of I in G. Since A is a Noetherian ring, the ideal I is already generated by a finite set of elements $b \cdot y - b(y)$. Hence there is a finite subset of B whose left fixer in G already coincides with H. Let B_0 be the smallest fully stable subalgebra of B that contains this finite subset. Clearly, B_0 is finitely generated as an F-algebra, and the left fixer of B_0 in G coincides with H.

Now let us first suppose that G is connected, i.e., that A is an integral domain. Let b be any element of B, and let x and y be elements of G whose restrictions to B_0 coincide. Then $x^{-1}y$ belongs to H, whence $x^{-1}y \cdot b = b$. Evaluating at x, we find that $x(b) = y(b)$. Hence it follows from Theorem 1.6 that b is purely inseparably algebraic over the field of fractions $[B_0]$ of B_0. Thus there is an exponent e such that b^{p^e} belongs to $[B_0]$, where p is the characteristic of F if that is not 0, and $p = 1$ if F is of characteristic 0. Since G separates the elements of A, the F-algebra homomorphisms $B_0 \to F$ separate the elements of B_0. Hence B_0 may be regarded as the algebra of all polynomial functions of the affine algebraic group $\mathcal{G}(B_0)$, and the restriction image of G in $\mathcal{G}(B_0)$ is algebraically dense in $\mathcal{G}(B_0)$. By Theorem 5.1, this restriction image is an algebraic subgroup of $\mathcal{G}(B_0)$. Hence the restriction image of G in $\mathcal{G}(B_0)$ coincides with $\mathcal{G}(B_0)$. If x is any element of G, we have $x \cdot (b^{p^e}) = (x \cdot b)^{p^e}$, whence we see that each $x \cdot (b^{p^e})$ lies in some fixed finite-dimensional subspace of $[B_0]$, depending only on b. By the result just above, these are *all* the transforms of b^{p^e} by the elements of $\mathcal{G}(B_0)$. Hence we may apply Lemma 6.3 to conclude that b^{p^e} actually belongs to B_0.

In particular, B is therefore contained in the integral closure of B_0 in

$[B]$. Clearly, $[A]$ is a finitely generated extension field of $[B_0]$, so that there is a finite transcendence basis $(t_1, \ldots, t_n)$ for $[A]$ over $[B_0]$, and $[A]$ is of finite dimension as a vector space over the field $[B_0](t_1, \ldots, t_n)$. A fortiori, $[B](t_1, \ldots, t_n)$ is of finite dimension over $[B_0](t_1, \ldots, t_n)$, whence there is a finite subset S of $[B]$ such that every element of $[B]$ is a $[B_0](t_1, \ldots, t_n)$-linear combination of elements of S. Since $[B]$ is algebraic over $[B_0]$, the set $(t_1, \ldots, t_n)$ is still algebraically free over $[B]$, and it follows that every element of $[B]$ is actually a $[B_0]$-linear combination of elements of S. Thus $[B]$ is a *finite* algebraic extension of $[B_0]$.

Now it follows from Theorem 1.12 that the integral closure of B_0 in $[B]$ is Noetherian as a B_0-module. Since B is contained in this integral closure, it is therefore finitely generated as a B_0-module. Since B_0 is finitely generated as an F-algebra, the same is therefore true for B, and we have now proved Theorem 6.4 in the case where G is connected.

In the general case, let G_1 denote the connected component of the neutral element in G, and let $A = A_1 + \cdots + A_n$ be the decomposition of A into a direct F-algebra sum that was obtained in Theorem 4.4. Then A_1 may be identified with the algebra of all polynomial functions of the affine algebraic group G_1. If f_1 is the identity element of A_1, then Bf_1 is a fully stable subalgebra of A_1, with respect to G_1. By what we have already proved, Bf_1 is therefore finitely generated as an F-algebra. For each i, if f_i is the identity element of A_i, we have $Bf_i = x_i \cdot (Bf_1)$, where x_i is some element of G. Hence each Bf_i is finitely generated as an F-algebra, and so is therefore the sum $Bf_1 + \cdots + Bf_n$. This sum evidently contains B, and we see immediately from Theorem 1.13 that B is therefore finitely generated as an F-algebra.

Finally, let ρ denote the restriction map $G \to \mathcal{G}(B)$. Since G separates the elements of B, we may view B as the algebra of all polynomial functions of the affine algebraic group $\mathcal{G}(B)$, and $\rho(G)$ is algebraically dense in $\mathcal{G}(B)$. The restriction of ρ to G_1 is a morphism of the connected affine algebraic group G_1 into $\mathcal{G}(B)$. By Theorem 5.1, $\rho(G_1)$ is therefore an algebraic subgroup of $\mathcal{G}(B)$. Now $\rho(G)$ is the union of a finite set of translates $\rho(x)\rho(G_1)$, and hence is an algebraic subset of $\mathcal{G}(B)$, i.e., the set of zeros in $\mathcal{G}(B)$ of some ideal of B. Since $\rho(G)$ is algebraically dense in $\mathcal{G}(B)$, we must therefore have $\rho(G) = \mathcal{G}(B)$. Our proof of Theorem 6.4 is now complete.

We observe that *the conclusion that B is finitely generated as an F-algebra holds without the assumption that F be algebraically closed.* In fact, let F' be an algebraic closure of F, and consider the extended structure $(G^{F'}, A \otimes F')$ of an affine algebraic group over F'. By Theorem 6.4, $B \otimes F'$ is finitely generated as an F'-algebra. It follows that there is a finitely generated F-subalgebra B_1 of B such that $B \subset B_1 \otimes F'$. Clearly, this implies $B = B_1$, so that B is finitely generated.

COROLLARY 6.5. *Let (G, A) be the structure of an affine algebraic group over an algebraically closed field F, and let H be a normal algebraic subgroup of G. Then the factor group G/H is an affine algebraic group, with the H-fixed part A^H of A as its algebra of polynomial functions. If $\rho: G \to K$ is a morphism of affine algebraic groups whose kernel contains H, then there is one and only one morphism $\rho^H: G/H \to K$ such that the composite of ρ^H with the restriction morphism $G \to \mathcal{G}(A^H) = G/H$ coincides with ρ.*

Proof: According to the notation employed earlier, A^H denotes the left H-fixed part of A. However, the normality of H in G is easily seen to imply that an element of A is left H-fixed if and only if it is right H-fixed. Thus A^H coincides with the right H-fixed part ${}^HA$ of A. In particular, it follows (or is seen directly) that A^H is a fully stable subalgebra of A. By Theorem 6.4, A^H is therefore finitely generated as an F-algebra, and the restriction map $G \to \mathcal{G}(A^H)$ is surjective. Moreover, since G separates the elements of A^H, we know that $(\mathcal{G}(A^H), A^H)$ is the structure of an affine algebraic group. By Theorem 6.2, the kernel of the restriction map $G \to \mathcal{G}(A^H)$ is precisely H, so that we may identify G/H with $\mathcal{G}(A^H)$.

Let $\rho: G \to K$ be as in the statement of the theorem. Since the kernel of ρ contains H, we have $\mathcal{A}(K) \circ \rho \subset A^H$. Hence we may define the required morphism $\rho^H: \mathcal{G}(A^H) \to K$ by $\rho^H(x)(f) = x(f \circ \rho)$ for every f in $\mathcal{A}(K)$. This completes the proof of Corollary 6.5.

In the case where the base field is of characteristic 0, there is an important addendum to the above results, which is as follows.

THEOREM 6.6. *Let (G, A) be the structure of an affine algebraic group over an algebraically closed field F of characteristic 0. Let B be a fully stable F-subalgebra of A whose fixer in G is trivial. Then $B = A$.*

Proof: Let ρ denote the restriction map $G \to \mathcal{G}(B)$. By Theorem 6.4, ρ is surjective, and the assumption on B means that ρ is injective. Let G_1 denote the connected component of the neutral element in G, and let $G_2, \ldots, G_p$ be all the other cosets of G_1 in G. As we have seen in proving Theorem 6.4, each $\rho(G_i)$ is an algebraic subset of the affine algebraic group $\mathscr{G}(B)$. Their union is $\mathscr{G}(B)$, and they are mutually disjoint, because ρ is injective. In fact, by Theorem 5.1, $\rho(G_1)$ is a connected algebraic subgroup of $\mathscr{G}(B)$, and the $\rho(G_i)$'s are the cosets of $\rho(G_1)$ in $\mathscr{G}(B)$. Hence $\rho(G_1)$ must be the connected component of the neutral element in $\mathscr{G}(B)$, by Theorem 4.1. Now let $B = B_1 + \cdots + B_p$ be the decomposition of B into a direct F-algebra sum that is obtained in Theorem 4.4. Let f_i denote the identity element of B_i. Since the $\rho(G_i)$'s are the cosets of $\mathscr{G}(B)_1$ in $\mathscr{G}(B)$, it is clear that the decomposition of A, as given in Theorem 4.4, is precisely $A = Af_1 + \cdots + Af_p$, in the present situation. We may identify Af_1 with the algebra of polynomial functions of G_1, and it is clear that Bf_1 is a fully stable subalgebra of Af_1 with respect to G_1, and that the fixer of Bf_1 in G_1 is trivial. A review of the first part of the proof of Theorem 6.4, with Bf_1 in the place of B_0, and with Af_1 in the place of B and A, noting that here the characteristic is 0, shows immediately that we must have $Bf_1 = Af_1$. Since each Af_i is a translate of Af_1, and since each Bf_i is the translate of Bf_1 by the same element of G, it follows that each Bf_i coincides with Af_i, whence $B = A$. This establishes Theorem 6.6.

Note that the essential content of Theorem 6.6 (bearing in mind the general fact that B is always finitely generated) is that, *for affine algebraic groups over an algebraically closed field of characteristic* 0, *the inverse of a bijective morphism is always a morphism, i.e., bijective morphisms are isomorphisms.*

EXERCISES

1. Let G be the group of all 2 by 2 matrices

$$x = \begin{pmatrix} \alpha(x) & \beta(x) \\ \gamma(x) & \delta(x) \end{pmatrix}$$

of determinant 1 with entries in a field F. Let A be the algebra of functions $F[\alpha, \beta, \gamma, \delta]$. Show that (G, A) is the structure of an affine algebraic group. Let H be the subgroup consisting of the matrices x for which $\gamma(x) = 0$. Find a set of semi-invariants defining H as an algebraic subgroup of G as in Theorem 6.1.

2. Let G, H, A be as in Exercise 1, and assume that F is an infinite field. Show that the left fixed part A^H of A is F. (Let f be an element of A^H, and write f as a polynomial $p(\alpha, \beta, \gamma, \delta)$. Now show that, for all triples (a, b, u) of elements of F in which $a \neq 0$, one has $p(a, b, ua, ub + a^{-1}) = p(1, 0, u, 1)$. In particular, $p(a, -1, 1, 0) =$

$p(1, 0, a^{-1}, 1)$, from which it follows that $p(1, 0, u, 1)$ is a fixed constant k for all u in F. Deduce from this that $(f - k)\alpha = 0$, whence $f = k$.)

3. Let F be an algebraically closed field of characteristic $p \neq 0$, and let F^* denote the multiplicative group of the non-zero elements of F. View F^* as an affine algebraic group, with $\mathscr{A}(F^*) = F[x,x^{-1}]$, where x is the identity map on F^*. Show that the pth power map $F^* \to F^*$ is a bijective morphism, but not an isomorphism of affine algebraic groups.

4. Let G be an affine algebraic group over an algebraically closed field of characteristic 0. Let H and K be algebraic subgroups of G such that $G = HK$ and K is normal in G. Prove that the natural morphism $H/(H \cap K) \to G/K$ is an isomorphism of affine algebraic groups.

5. Show that the result of Exercise 4 is not generally valid over an algebraically closed field F of characteristic $p \neq 0$ by examining the following example. Let G be the direct product $F^* \times F^*$, with F^* as in Exercise 3. Let K be the subgroup consisting of the elements (a, a^p), and let H be the subgroup of the elements $(a, 1)$.

7. THE LIE ALGEBRA

Let (G, A) be the structure of an affine algebraic group over an arbitrary field F. By a *differentiation* of A we mean an F-linear map $\delta : A \to F$ such that $\delta(ab) = c(a)\delta(b) + \delta(a)c(b)$ for all elements a and b of A, where c denotes the co-unit of A. In the geometric language, this means that δ is a tangent to G at the neutral element.

Recall from Section 2 that there is a linear isomorphism

$$\sigma \to (i_A \otimes \sigma) \circ \gamma$$

of the dual space A° of A onto the space of all proper linear endomorphisms of A, a linear endomorphism e of A being called proper if $\gamma \circ e = (i_A \otimes e) \circ \gamma$ or, equivalently, if e commutes with every right translation effected by an element of G on A. The inverse of this isomorphism is the map $e \to c \circ e$. Let us abbreviate $(i_A \otimes \sigma) \circ \gamma$ by σ^*, and recall that, if σ and τ are arbitrary elements of A°, we have $\sigma^* \circ \tau^* = (\sigma \otimes \tau) \circ \gamma$.

One verifies immediately that an element δ of A° is a differentiation if and only if δ^* is a *derivation* of A, i.e., if and only if

$$\delta^*(ab) = a\delta^*(b) + \delta^*(a)b$$

for all elements a and b of A. Thus our above isomorphism sends the space of all differentiations of A onto the space of all proper derivations of A. The second space is a Lie algebra under the usual Lie composition

$$(e_1, e_2) \to [e_1, e_2] = e_1 \circ e_2 - e_2 \circ e_1 .$$

Our isomorphism transports this to a Lie composition on the space of the differentiations of A given by

$$(\delta_1, \delta_2) \to [\delta_1, \delta_2] = (\delta_1 \otimes \delta_2 - \delta_2 \otimes \delta_1) \circ \gamma .$$

This Lie algebra, whose elements are the differentiations of A, is called the Lie algebra of G, and is denoted by $\mathscr{L}(G)$. Since A is finitely generated as an F-algebra, it is clear that $\mathscr{L}(G)$ is of finite dimension as an F-space.

Let $\rho : G \to H$ be a morphism of affine algebraic groups. For δ in $\mathscr{L}(G)$, define the linear functional $\rho^\circ(\delta)$ on $A(H)$ by $\rho^\circ(\delta)(f) = \delta(f \circ \rho)$. It is seen directly from the definitions that ρ° is a Lie algebra homomorphism $\mathscr{L}(G) \to \mathscr{L}(H)$. This Lie algebra homomorphism ρ° is called the *differential* of ρ. Evidently, this makes $\mathscr{L}$ a functor from the category of affine algebraic groups over F to the category of Lie algebras over F. When ρ is the canonical injection of an algebraic subgroup, ρ° is injective, and we identify $\mathscr{L}(G)$ with $\rho^\circ(\mathscr{L}(G))$.

In dealing with Lie algebras of affine algebraic groups, nothing is lost by restricting one's attention to connected groups, as is seen from the following consideration. Let G be an affine algebraic group, and let G_1 be the connected component of the neutral element in G. Let ρ be the canonical injection morphism $G_1 \to G$. We have $\mathscr{A}(G_1) = A/P$, where $A = \mathscr{A}(G)$ and P is the annihilator of G_1 in A. The map $f \to f \circ \rho$ of $\mathscr{A}(G)$ into $\mathscr{A}(G_1)$ is the canonical homomorphism $A \to A/P$. From the fact that this map is surjective, it follows immediately that ρ° is injective. Now recall from the beginning of Section 4 that there is an element q in A such that $c(q) \neq 0$ and P is the annihilator of q in A. Let δ be any element of $\mathscr{L}(G)$, and let p be an element of P. Then $0 = \delta(pq) = c(p)\delta(q) + c(q)\delta(p)$. Since $c(p) = 0$ and $c(q) \neq 0$, this give $\delta(p) = 0$. Hence it is clear that δ induces a differentiation β of A/P in the natural fashion. Evidently, $\rho^\circ(\beta) = \delta$. Thus we conclude that *the differential of the injection $G_1 \to G$ is a Lie algebra isomorphism of $\mathscr{L}(G_1)$ onto $\mathscr{L}(G)$.*

A similar working principle is derived from the fact that the effect on the Lie algebra of an extension of the base field is essentially trivial. We proceed to make this precise. Let (G, A) be the structure of an affine algebraic group over an arbitrary field F. Let L be a field containing F, and consider the extended affine algebraic group structure $(G^L, A \otimes L)$. If δ is an element of $\mathscr{L}(G)$, then the L-linear extension of δ to an L-linear map $A \otimes L \to L$ is evidently an element of $\mathscr{L}(G^L)$. Thus we have a homomorphism of L-Lie algebras $\mathscr{L}(G) \otimes L \to \mathscr{L}(G^L)$, which we call the canonical homomorphism.

PROPOSITION 7.1. *Let (G, A) be the structure of an affine algebraic group over an arbitrary field F, and let L be a field containing F. Then the canonical Lie algebra homomorphism $\mathscr{L}(G) \otimes L \to \mathscr{L}(G^L)$ is an isomorphism.*

Proof: Clearly, the canonical homomorphism is injective. In order to see that it is surjective, consider an arbitrary element δ of $\mathscr{L}(G^L)$. Choose an F-basis (b_α) of L. For every element a of A (identified with its image in $A \otimes L$), write $\delta(a) = \sum_\alpha \delta_\alpha(a) b_\alpha$, with each $\delta_\alpha(a)$ in F. This defines, for each index α, a linear map $\delta_\alpha : A \to F$, and we see immediately that δ_α belongs to $\mathscr{L}(G)$. Now A has a finite system $(a_1, \ldots, a_n)$ of F-algebra generators, which is also a system of L-algebra generators of $A \otimes L$. The elements $\delta(a_i)$, when written out as above, involve altogether only a finite set of b_α's (for a given a, $\delta_\alpha(a) = 0$ for all but a finite set of indices α): $b_1, \ldots, b_m$. The canonical image in $\mathscr{L}(G^L)$ of the element $\sum_{i=1}^m \delta_i \otimes b_i$ of $\mathscr{L}(G) \otimes L$ maps each generator a_j onto $\delta(a_j)$, whence it must coincide with δ. This establishes Proposition 7.1.

THEOREM 7.2. *Let (G, A) be the structure of an affine algebraic group over an arbitrary field F. Let $\mathscr{D}(A)$ denote the A-module of all F-algebra derivations of A. Then the A-module homomorphism $A \otimes \mathscr{L}(G) \to \mathscr{D}(A)$ that sends each element of $\mathscr{L}(G)$ onto the corresponding proper derivation of A is an isomorphism of A-modules.*

Proof: Let $(\sigma_1, \ldots, \sigma_n)$ be an F-basis of $\mathscr{L}(G)$, and write τ_i for the proper derivation $\sigma_i{}^*$ of A that corresponds to σ_i. Then every element of $A \otimes \mathscr{L}(G)$ can be written in the form $\sum_{i=1}^n a_i \otimes \sigma_i$, with uniquely determined elements a_i of A, and its image in $\mathscr{D}(A)$ is $\sum_{i=1}^n a_i \tau_i$. If this image is 0, then we have, for every a in A and every x in G, $\sum_{i=1}^n a_i \tau_i(a \cdot x^{-1}) = 0$. Applying the right translation effected by x to this, and noting that τ_i is proper, we obtain $\sum_{i=1}^n (a_i \cdot x)\tau_i(a) = 0$. Evaluating this at the neutral element of G, we obtain

$$\sum_{i=1}^n a_i(x)\sigma_i(a) = 0 .$$

Since this holds for every a in A, and since the differentiations σ_i are linearly independent over F, we must have $a_i(x) = 0$ for each index i. Since this holds for all elements x of G, we conclude that each a_i is 0. Thus we have shown that our A-module homomorphism is injective.

In proving that this homomorphism is also surjective, it is convenient to use the following G-module structure on $\mathscr{D}(A)$. Let δ be an element of $\mathscr{D}(A)$, and let x be an element of G. We define the element $x \cdot \delta$ of $\mathscr{D}(A)$ by

$$(x \cdot \delta)(a) = \delta(a \cdot x) \cdot x^{-1} .$$

One verifies directly that $x \cdot \delta$ is actually an F-algebra derivation of A, and that our definition makes $\mathscr{D}(A)$ into a G-module. Clearly, the G-fixed part of $\mathscr{D}(A)$ is precisely the space of the proper derivations of A. From the fact that A is locally finite with respect to the action of G by right translations, and that an element of $\mathscr{D}(A)$ is already determined by its effect on some finite system of F-algebra generators of A, we see that $\mathscr{D}(A)$ is locally finite as a G-module, and that the associated representative functions belong to A.

Now let δ be any element of $\mathscr{D}(A)$, and let c denote the co-unit of A. Then, for every element x of G, the composite $c \circ (x^{-1} \cdot \delta)$ is an element of $\mathscr{L}(G)$, so that there are uniquely determined elements $a_i(x)$ of F such that $c \circ (x^{-1} \cdot \delta) = \sum_{i=1}^{n} a_i(x)\sigma_i$. By our last remark above, the functions a_i so defined are elements of A. We claim that $\delta = \sum_{i=1}^{n} a_i\tau_i$. In order to verify this, let a be an element of A, and x an element of G. Then we have

$$\left(\sum_{i=1}^{n} a_i\tau_i\right)(a)(x) = \sum_{i=1}^{n} a_i(x)\tau_i(a)(x) = \sum_{i=1}^{n} a_i(x)\sigma_i(a \cdot x)$$
$$= \big(c \circ (x^{-1} \cdot \delta)\big)(a \cdot x) = c\big(\delta(a) \cdot x\big) = \delta(a)(x),$$

which proves what we claimed above. Thus our arbitrarily given element δ of $\mathscr{D}(A)$ is the image of the element $\sum_{i=1}^{n} a_i \otimes \sigma_i$ of $A \otimes \mathscr{L}(G)$. Our proof of Theorem 7.2 is now complete.

COROLLARY 7.3. *Let (G, A) be as in Theorem 7.2, and let $[A]$ denote the total ring of fractions of A. By tensoring the isomorphism of Theorem 7.2 with $[A]$, relative to A, one obtains an isomorphism of $[A]$-modules*

$$[A] \otimes \mathscr{L}(G) \to \mathscr{D}([A]).$$

Proof: Evidently, we may identify $[A] \otimes_A \big(A \otimes (\mathscr{L}(G)\big)$ with $[A] \otimes \mathscr{L}(G)$, so that tensoring the isomorphism of Theorem 7.2 with $[A]$ yields an isomorphism of $[A]$-modules $[A] \otimes \mathscr{L}(G) \to [A] \otimes_A \mathscr{D}(A)$. Consider the evident $[A]$-module homomorphism of $[A] \otimes_A \mathscr{D}(A)$ into the $[A]$-module $\mathscr{D}(A, [A])$ of all F-algebra derivations of A into $[A]$. Every element of $[A] \otimes_A \mathscr{D}(A)$ can be written in the form $\sum_i (a_i/d) \otimes \delta_i$, with each δ_i in $\mathscr{D}(A)$, each a_i in A, and d a non-zerodivisor in A. If the image of this element in $\mathscr{D}(A, [A])$ is 0, then so is its multiple by d. Noting that the canonical map $A \to [A]$ is injective, we conclude that the element $\sum_i a_i\delta_i$ of $\mathscr{D}(A)$ is 0. Clearly, this implies that our given element of $[A] \otimes_A \mathscr{D}(A)$ is 0. Thus the natural map $[A] \otimes_A \mathscr{D}(A) \to \mathscr{D}(A, [A])$ is injective. From the fact that A is finitely generated as an F-algebra, we see that, given any element τ of $\mathscr{D}(A, [A])$, there is a non-zerodivisor d in A such that $d\tau$ maps A into the image of A in $[A]$. Hence we may regard $d\tau$ as an element of $\mathscr{D}(A)$, and τ is then the image of the element $(1/d) \otimes (d\tau)$ of $[A] \otimes_A \mathscr{D}(A)$. Thus we may identify $[A] \otimes_A \mathscr{D}(A)$ with $\mathscr{D}(A, [A])$. Finally, the usual formula for differentiating a fraction, $\tau(a/d) = \big(\tau(a)d - a\tau(d)\big)/d^2$, applies to yield an identification of $\mathscr{D}(A, [A])$ with the $[A]$-module $\mathscr{D}([A])$ of all F-algebra derivations of $[A]$. This completes the proof of Corollary 7.3.

THEOREM 7.4. *Let (G, A) be the structure of a connected affine algebraic group over an arbitrary field F. Then the F-dimension of $\mathscr{L}(G)$ is equal to the transcendence degree of the field $[A]$ over F (i.e., it is equal to the dimension of G as an affine algebraic variety).*

Proof: By Corollary 7.3, the F-dimension of $\mathscr{L}(G)$ is equal to the $[A]$-dimension of $\mathscr{D}([A])$. Hence it suffices to show that the $[A]$-dimension of $\mathscr{D}([A])$ is equal to the transcendence degree of $[A]$ over F.

An elementary argument concerning spaces of maps of a set (here $[A]$) into a field (here $[A]$) shows that we can find an $[A]$-basis $(\delta_1, \ldots, \delta_m)$ of $\mathscr{D}([A])$, and a corresponding subset $(u_1, \ldots, u_m)$ of $[A]$, such that $\delta_i(u_j)$ is equal to 1 or 0 according to whether $i = j$ or $i \neq j$. Let K denote the subfield $F(u_1, \ldots, u_m)$ of $[A]$. First, we show that $[A]$ is separably algebraic over K. We know that $[A]$ is finitely generated as an extension field of F. Hence there is a finite subset $(v_1, \ldots, v_n)$ of $[A]$ such that $[A] = K(v_1, \ldots, v_n)$. Let k be the smallest index such that $[A]$ is separably algebraic over the field generated by K and the v_i's with $i \leqq k$. If $k > 0$, then either v_k is transcendental over $K(v_1, \ldots, v_{k-1})$, or v_k is algebraic and inseparable over $K(v_1, \ldots, v_{k-1})$, in which case F is of characteristic $p \neq 0$ and v_k does not belong to $K(v_1, \ldots, v_{k-1}, v_k{}^p)$. In each of these two cases, it is easy to see that there is a non-zero derivation of $K(v_1, \ldots, v_k)$ that annihilates $K(v_1, \ldots, v_{k-1})$. Since $[A]$ is separably algebraic over $K(v_1, \ldots, v_k)$, this derivation extends to a derivation of $[A]$. On the other hand, every element of $\mathscr{D}([A])$ is an $[A]$-linear combination of the δ_i's, whence it is clear that a non-zero element of $\mathscr{D}([A])$ cannot annihilate K. Thus we conclude that we must have $k = 0$, which means that $[A]$ is separably algebraic over K.

Our theorem will therefore be proved as soon as we have shown that the elements u_j are algebraically independent over F. Suppose this is not the case, and let f be a non-zero polynomial of the least possible degree in m variables $x_1, \ldots, x_m$, with coefficients in F, such that $f(u_1, \ldots, u_m) = 0$. Applying δ_i to this, we find that $f_i(u_1, \ldots, u_m) = 0$, where f_i denotes the formal derivative of f with respect to x_1. By the minimality of f, this implies that each f_i is 0. Hence F has non-zero characteristic p, and there is a polynomial g in the x_j's, with coefficients in F, such that

$$f(x_1, \ldots, x_m) = g(x_1{}^p, \ldots, x_m{}^p) = \sum_k s_k y_k{}^p,$$

where each s_k is in F, and the y_k's are certain monomials formed with the x_j's. Let w_k be the element of $[A]$ that is obtained from y_k by replacing each x_j with u_j. Then we have $\sum_k s_k w_k{}^p = 0$. Let L be an algebraic closure of F. We know that $A \otimes L$ is an integral domain (Prop. 4.2). Since L is algebraic over F, the field of fractions of $A \otimes L$ is $[A] \otimes L$. Now choose elements t_j in L such that $t_j{}^p = s_j$. Viewing the above as a relation in $[A] \otimes L$, we have

$$0 = \sum_k s_k w_k{}^p = \Big(\sum_k w_k \otimes t_k\Big)^p$$

whence $\sum_k w_k \otimes t_k = 0$. By the minimality of f, the w_k's are linearly independent over F. Hence each t_k must be 0, so that each s_k is 0. This gives the contradiction $f = 0$, and our proof of Theorem 7.4 is now complete.

THEOREM 7.5. *Let (G, A) and (H, B) be structures of affine algebraic groups over an arbitrary field F, and let $\rho : G \to H$ be a morphism of affine algebraic groups. Suppose that G is connected, that $\rho(G)$ is algebraically dense in H, and that $[A]$ is separably generated as an extension field of $[B \circ \rho]$. Then the differential $\rho^\circ : \mathscr{L}(G) \to \mathscr{L}(H)$ of ρ is surjective.*

Proof: Since $\rho(G)$ is algebraically dense in H, the map $b \to b \circ \rho$ of B into A is injective, so that we may identify B with $B \circ \rho$. Let τ be any element of $\mathscr{L}(H)$, and let τ^* denote the corresponding proper derivation of B. Clearly, τ^* extends uniquely to a derivation of the field $[B]$. Since $[A]$ is a separably generated extension field of $[B]$, this derivation can be extended further to a derivation δ of $[A]$. Since A is finitely generated as an F-algebra, there is a non-zero element a in A such that $a\delta$ stabilizes A. The G-module structure of $\mathscr{D}(A)$ that we used in the proof of Theorem 7.2 has an evident natural extension to a G-module structure of $\mathscr{D}([A])$. Choose an element x in G such that $a(x^{-1}) \neq 0$, and consider the transformed derivation $x \cdot (a\delta)$ of $[A]$. Evidently, this still stabilizes A. Moreover, we have $x \cdot (a\delta) = (a \cdot x^{-1})(x \cdot \delta)$. Clearly, $x \cdot \delta$ coincides with $x \cdot \tau^* = \tau^*$ on B. Thus, replacing δ with $x \cdot \delta$ if necessary, we achieve $c(a) \neq 0$, where c denotes the co-unit of A. This shows that $\delta(A)$ lies in the specialization ring of c in $[A]$, so that the composite $c \circ \delta$ is definable on A. Clearly, this is a differentiation $A \to F$, i.e., it is an element, σ say, of $\mathscr{L}(G)$. The restriction of σ to B coincides with $c \circ \tau^* = \tau$, which means that $\rho^\circ(\sigma) = \tau$. This establishes Theorem 7.5.

If the field F is of characteristic 0, then the assumption of separable generation is automatically satisfied, and we may express Theorem 7.5 by saying that *if $\rho : G \to H$ is a morphism of (not necessarily connected) affine algebraic groups over a field of characteristic* 0 *then the differential of ρ is surjective from $\mathscr{L}(G)$ to the Lie algebra of the algebraic hull of $\rho(G)$ in H.*

EXERCISES

1. Let F be a field of characteristic $p \neq 0$, let F^* be the multiplicative group of the non-zero elements of F, viewed as an affine algebraic group as in Exercise 3 of Section 6. Let π: $F^* \to F^*$ be the pth power map. Show that the differential of π is the 0-map.
2. Let F be an arbitrary field, and let G be the group of pairs (a, u), where a ranges over F and u ranges over F^*, the product being defined by
$$(a_1, u_1)(a_2, u_2) = (a_1 + u_1a_2, u_1u_2)\,.$$
Define the functions α and β on G by
$$\alpha(a, u) = a, \qquad \beta(a, u) = u\,.$$
Let A be the algebra of functions $F[\alpha, \beta, \beta^{-1}]$. Show that (G, A) is the structure of an affine algebraic group, and determine the Lie algebra of G.

8. LIE SUBALGEBRAS AND ALGEBRAIC SUBGROUPS

In order not to disrupt continuity later on, we begin with two purely technical lemmas.

LEMMA 8.1. *Let K be a field of characteristic* 0, *and let $K[[t]]$ denote the ring of integral power series in a variable t over K. Let $(a_1, \ldots, a_q)$ be a subset of K that is linearly indepenedent over the field of rational numbers. Then the elements t, $\exp(a_1 t), \ldots, \exp(a_q t)$ of $K[[t]]$ are algebraically independent over K.*

Proof: Let $x_0, \ldots, x_q$ denote independent variables over K, and suppose that f is an element of the polynomial algebra $K[x_0, \ldots, x_q]$ such that $f(t, \exp(a_1 t), \ldots, \exp(a_q t)) = 0$. Write $f_{(e_0, \ldots, e_q)}$ for the coefficient of $x_0^{e_0} \cdots x_q^{e_q}$ in $f(x_0, \ldots, x_q)$. Then we have

$$\sum f_{(e_0, \ldots, e_q)} t^{e_0} \exp((e_1 a_1 + \cdots + e_q a_q)t) = 0,$$

and the assumption on the set $(a_1, \ldots, a_q)$ means that no two of the linear combinations $e_1 a_1 + \cdots + e_q a_q$ are equal. Hence Lemma 8.1 will be established as soon as we have proved the following: if $b_1, \ldots, b_n$ are n distinct elements of K, then the n power series $\exp(b_1 t), \ldots, \exp(b_n t)$ are linearly independent over the polynomial ring $K[t]$.

Suppose that this result has been proved for $(n-1)$-tuples. Let $p_i(t)$ be elements of $K[t]$ such that $\sum_{i=1}^{n} p_i(t) \exp(b_i t) = 0$. Multiplying by $\exp(-b_n t)$, we obtain

$$p_n(t) + \sum_{i=1}^{n-1} p_i(t) \exp((b_i - b_n)t) = 0.$$

Differentiating k times with respect to t, we obtain

$$p_n^{(k)}(t) + \sum_{i=1}^{n-1} p_{i,k}(t) \exp((b_i - b_n)t) = 0 ,$$

where the $p_{i,k}(t)$'s are elements of $K[t]$ that are determined by the recursive relations $p_{i,0}(t) = p_i(t)$, and

$$p_{i,h+1}(t) = p_{i,h}^{(1)}(t) + (b_i - b_n)p_{i,h}(t) ,$$

the parenthetical exponents denoting orders of derivatives. If k is large enough, we have $p_n^{(k)}(t) = 0$, and then the inductive hypothesis gives $p_{i,k}(t) = 0$ for each i. Hence

$$p_{i,k-1}^{(1)}(t) + (b_i - b_n)p_{i,k-1}(t) = 0 ,$$

from which it is readily seen that $p_{i,k-1}(t) = 0$ for each i. We can evidently repeat this last argument until we obtain $p_i(t) = 0$ for each i. This completes the proof of Lemma 8.1.

LEMMA 8.2 *Let K be a field, and let $(u_1, \ldots, u_r)$ be independent variables over K. Let S be a finite group of field automorphisms of $K(u_1, \ldots, u_r)$ that stabilizes K, is represented faithfully on K, and stabilizes the multiplicative group generated by $(u_1, \ldots, u_r)$. Then the S-fixed part of $K(u_1, \ldots, u_r)$ is contained in a finitely generated purely transcendental extension field of the S-fixed part of K.*

Proof: Let K^S denote the S-fixed part of K, and choose a K^S-basis $(k_1, \ldots, k_n)$ of K (so that n is the order of S). Let v_{ij} $(i = 1, \ldots, r$, and $j = 1, \ldots, n)$ be independent variables over K, and consider the field $K(v_{11}, \ldots, v_{rn})$. The multiplicative group U that is generated by $(u_1, \ldots, u_r)$ in $K(u_1, \ldots, u_r)$ may evidently be identified with the free abelian group based on the set $(u_1, \ldots, u_r)$, so that there is one and only one group homomorphism π of U into the multiplicative group of $K(v_{11}, \ldots, v_{rn})$ such that

$$\pi(u_i) = \sum_{j=1}^{n} k_j v_{ij}$$

for each i. The action of S on K extends uniquely to an action of S by field automorphisms of $K(v_{11}, \ldots, v_{rn})$ leaving the elements v_{ij} fixed. For each element s of S, define the group homomorphism $\pi_s : U \to K(v_{11}, \ldots, v_{rn})$ by $\pi_s(u) = s(\pi(s^{-1}(u)))$. Let η denote the (value-wise) product of all these homomorphisms π_s. Then η is still a group homomorphism, and $\eta(s(u)) = s(\eta(u))$ for every s in S and every u in U.

From elementary Galois theory, we know that the K-space spanned by the automorphisms s' of K corresponding to the elements s of S is the space of all K^S-linear endomorphisms of K. Hence, for each $q = 1, \ldots, n$, there are elements $c_{s,q}$ in K such that the endomorphism $\sum_{s \in S} c_{s,q} s'$ of K sends k_q onto 1 and annihilates each k_j with $j \neq q$. From the definition of π, we therefore obtain

$$\sum_{s \in S} c_{s,q} s(\pi(u_i)) = v_{iq} .$$

Thus each v_{iq} is a K-linear combination of the rn elements $s(\pi(u_i))$. It follows that these last rn elements are algebraically independent over K.

Let N_s denote the multiplicative group generated in $K(v_{11}, \ldots, v_{rn})$ by the elements $s(\pi(u_i))$, with $i = 1, \ldots, r$, and let N denote the group generated by all the N_s's. The algebraic independence of the elements $s(\pi(u_i))$ implies that N is the direct product of the groups N_s, that each π_s is an isomorphism of U onto N_s, and hence that their product η is injective. Since the elements of N are linearly independent over K, the K-linear extension of η is an injective K-algebra homomorphism of $K[U]$ into $K(v_{11}, \ldots, v_{rn})$, and therefore extends further to a field injection of $K(u_1, \ldots, u_r)$, which is the field of fractions of $K[U]$, into $K(v_{11}, \ldots, v_{rn})$. Let us denote this injection by τ. Clearly, τ extends the canonical injection $K \to K(v_{11}, \ldots, v_{rn})$. Moreover, since $\eta(s(u)) = s(\eta(u))$ for every s in S and every u in U, we have $\tau(s(x)) = s(\tau(x))$ for every s in S and every x in $K(u_1, \ldots, u_r)$. It follows that τ sends the S-fixed part of $K(u_1, \ldots, u_r)$ into the S-fixed part of $K(v_{11}, \ldots, v_{rn})$, which is evidently $K^S(v_{11}, \ldots, v_{rn})$. Clearly, this yields Lemma 8.2.

Now let (G, A) be the structure of an affine algebraic group over an arbitrary field F. Let σ be an element of $\mathscr{L}(G)$, and let σ^* be the corresponding proper derivation of A. Let c denote the co-unit of A. Let I_σ denote the F-subspace of A consisting of all elements a with the property that $c(\sigma^{*n}(a)) = 0$ for every non-negative exponent n (σ^{*0} stands for the identity map on A). One sees immediately that I_σ is an ideal of A. We shall show that I_σ is actually a Hopf ideal of A, in the sense of Section 3. Since we evidently have $c(I_\sigma) = (0)$, it suffices to show that, if γ is the comultiplication of A, we have $\gamma(I_\sigma) \subset I_\sigma \otimes A + A \otimes I_\sigma$. Observe that if p and q are non-negative integers such that $p + q = n$, then $c \circ \sigma^{*n} = (c \circ \sigma^{*p} \otimes c \circ \sigma^{*q}) \circ \gamma$. Hence $\gamma(I_\sigma)$ lies in the intersection of the family of the kernels in $A \otimes A$ of the linear functionals $c \circ \sigma^{*p} \otimes c \circ \sigma^{*q}$, with p and q ranging over all non-negative integers. An argument of elementary linear algebra shows that this intersection is precisely $I_\sigma \otimes A + A \otimes I_\sigma$, which proves that I_σ is indeed a Hopf ideal of A.

Let G_σ denote the annihilator of I_σ in G. Since I_σ is a Hopf ideal, we know from Proposition 3.1 that G_σ is a subgroup of G, and hence an algebraic subgroup of G. Suppose that H is any algebraic subgroup of G such that σ belongs to $\mathscr{L}(H)$. Let J be the annihilator of H in A. Then J is a Hopf ideal of A, and $\sigma(J) = (0)$. Recalling that $\sigma^* = (i_A \otimes \sigma) \circ \gamma$, we see immediately from these two facts that $\sigma^*(J) \subset J$, and hence that $J \subset I_\sigma$. Hence $G_\sigma \subset H$.

In the case where F is of characteristic $p \neq 0$, we have $(a - c(a))^p \in I_\sigma$ for every element a of A, whence G_σ is trivial. If F is of characteristic 0, then G_σ is of significance, by virtue of the following theorem.

THEOREM 8.3 *Let (G, A) be the structure of an affine algebraic group over the field F of characteristic 0, let σ be an element of $\mathscr{L}(G)$, and let G_σ be the corresponding algebraic subgroup of G, as defined above. Then G_σ is contained in every algebraic subgroup of G whose Lie algebra contains σ. Moreover, G_σ is connected and σ belongs to $\mathscr{L}(G_\sigma)$.*

Proof: It remains only to prove that G_σ is connected, and that σ belongs to $\mathscr{L}(G_\sigma)$. First, let us show that I_σ is a prime ideal. Let a and b be elements of A that do not belong to I_σ. Let p be the smallest exponent such that $c(\sigma^{*p}(a)) \neq 0$, and let q be the smallest exponent such that $c(\sigma^{*q}(b)) \neq 0$. Since σ^* is a derivation, we have

$$c(\sigma^{*p+q}(ab)) = \sum_{u+v=p+q} \frac{(p+q)!}{u!\,v!}\, c(\sigma^{*u}(a)\sigma^{*v}(b))$$

By the definitions of p and q, this reduces to the single non-zero term with $u = p$ and $v = q$. Hence ab does not belong to I_σ, and we have shown that I_σ is a prime ideal.

Now it is clear that our theorem will be proved as soon as we have shown that the annihilator of G_σ in A coincides with I_σ. Using the fact that I_σ is a prime ideal, we may conclude this immediately from Theorem 1.5, in the case where F is algebraically closed. In the general case, we shall be able to replace the appeal to Theorem 1.5 with an explicit specialization argument, once we have shown that the integral domain A/I_σ is contained in a finitely generated purely transcendental field extension of F.

Let us choose a finite-dimensional left stable F-subspace V of A that generates A as an F-algebra. We know from Section 2 that V is stable under σ^*. Let us now choose a finite Galois extension K of F such that the characteristic polynomial of the restriction of σ^* to V splits into linear factors in the polynomial ring over K. We identify σ with its canonical image in $\mathscr{L}(G^K)$ (see Proposition 7.1), i. e., with its K-linear extension to a differentiation $A \otimes K \to K$. Compatibly, we identify σ^* with its K-linear extension to a proper derivation of $A \otimes K$. Let $c_1, \ldots, c_n$ be all the characteristic roots of the restriction of σ^* to $V \otimes K$. These are, of course, the characteristic roots of the restriction of σ^* to V, and thus lie in K. Hence $V \otimes K$ is the direct sum of σ^*-stable K-subspaces $V_1, \ldots, V_n$, such that each V_i is annihilated by some power of $\sigma^* - c_i$.

Let t be a variable over K, and consider the K-algebra homomorphism $\exp(t\sigma)\colon A \otimes K \to K[[t]]$, where we have written $\exp(t\sigma)$ for

$$c \circ \exp(t\sigma^*) = \sum_{i \geqq 0} \frac{t^i}{i!} c \circ \sigma^{*i} .$$

Clearly, the kernel of $\exp(t\sigma)$ is precisely $I_\sigma \otimes K$. Let us write μ_i for the restriction of $\sigma^* - c_i$ to V_i. Then the restriction to V_i of $\exp(t\sigma^*)$ is $\exp(tc_i)\exp(t\mu_i)$. Moreover, there is an exponent d_i such that $\mu_i^{d_i+1} = 0$, whence

$$\exp(tc_i)\exp(t\mu_i) = \exp(tc_i) \sum_{j=1}^{d_i} \frac{t^j}{j!} \mu_i{}^j .$$

It follows immediately from this that $\exp(t\sigma)(V_i) \subset K[t, \exp(tc_i)]$. Since V generates A as a K-algebra, we have therefore

$$\exp(t\sigma)(A \otimes K) \subset K[t, \exp(tc_1), \ldots, \exp(tc_n)] .$$

Although this conclusion suffices for the present purpose, we remark that each V_i contains an element v_i such that $\sigma^*(v_i) = c_i v_i$. Moreover, replacing V with a suitable translate $V \cdot x$ if necessary, we can arrange to have, furthermore, $c(v_i) = 1$, in which case $\exp(t\sigma)(v_i) = \exp(tc_i)$. Hence $K[\exp(tc_1), \ldots, \exp(tc_n)] \subset \exp(t\sigma)(A \otimes K)$.

Since the additive subgroup of K that is generated by the c_i's has no torsion, it is the free abelian group based on a set $(a_1, \ldots, a_q)$ of in tegral linear combinations of the c_i's. The field of fractions of $K[t, \exp(tc_1), \ldots, \exp(tc_n)]$ is therefore $K(t, \exp(ta_1), \ldots, \exp(ta_q))$, and we have from Lemma 8.1 that this field is a purely transcendental extension of K, of transcendence degree $q + 1$.

Let S denote the Galois group of K over F, and let S act coefficient-wise on $K[[t]]$. Since the characteristic polynomial of the restriction of σ^* to $V \otimes K$ has its coefficients in F, the set $(c_1, \ldots, c_n)$ of characteristic roots is S-stable. Hence S stabilizes the multiplicative group generated by $(t, \exp(ta_1), \ldots, \exp(ta_q))$ in $K[[t]]$. The action of S on $K[[t]]$ extends uniquely to an action of S by field automorphisms on the field of fractions of $K[[t]]$, and clearly the subfield $K(t, \exp(ta_1), \ldots, \exp(ta_q))$ is S-stable. Since $\exp(t\sigma)(A) \subset F[[t]]$, it is contained in the S-fixed part of $K(t, \exp(ta_1), \ldots, \exp(ta_q))$. Now we can apply Lemma 8.2 to conclude that $\exp(t\sigma)(A)$ is contained in a finitely generated purely transcendental extension of F. Since the kernel of $\exp(t\sigma)$ in A is I_σ, we have therefore $A/I_\sigma \subset F(x_1, \ldots, x_m)$, where $(x_1, \ldots, x_m)$ is algebraically independent over F.

Now let J denote the annihilator of G_σ in A. Clearly, $I_\sigma \subset J$. Suppose that $J \neq I_\sigma$, and let b be an element of J that does not belong to I_σ. Let $a \to a'$ denote the canonical homomorphism $A \to A/I_\sigma$. Let $(p_1, \ldots, p_k)$ be a set of F-algebra generators of A, so ordered that $p_1, \ldots, p_j$ are those p_i's which do not belong to I_σ. Then, for every a in A, the image a' is a polynomial in $p_1', \ldots, p_j'$, with coefficients in F. Write $p_i' = f_i/g_i$ $(i = 1, \ldots, j)$, and $b' = f/g$, where the f_i's, the g_i's, f and g are non-zero elements of $F[x_1, \ldots, x_m]$. Since F is infinite, we can find elements $r_1, \ldots, r_m$ in F such that

$$(g_1 \cdots g_m g f)(r_1, \ldots, r_m) \neq 0 .$$

Then it is clear that the specialization $x_i \to r_i$ defines an F-algebra homomorphism $A/I_\sigma \to F$ that does not annihilate b'. This may be viewed as an F-algebra homomorphism $A \to F$ that annihilates I_σ, but does not annihilate b. Thus we have an element of G_σ that does not annihilate b, contradicting the fact that b belongs to J. Hence we must have $J = I_\sigma$, and our proof of Theorem 8.3 is now complete.

PROPOSITION 8.4 *In the notation of Theorem* 8.3, *with F of characteristic* 0, *let q denote the rank of the additive group generated by the characteristic roots of* σ^*. *Then the dimension of* $\mathscr{L}(G_\sigma)$ *is equal to q or* $q + 1$ *according to whether* σ^* *is semisimple or not.*

Proof: Clearly, the rank q described here is precisely the index q that occured in our proof of Theorem 8.3, because the additive group generated by the characteristic roots $c_1, \ldots, c_n$ used in that proof coincides with the additive group generated by *all* the characteristic roots of σ^*, owing to the fact that the space V used in proving Theorem 8.3 generates A as an F-algebra, and that σ^* is a derivation. By Proposition 7.1, the F-dimension of $\mathscr{L}(G_\sigma)$ is equal to the K-dimension of $\mathscr{L}(G_\sigma^K)$, where K is the field in the proof of Theorem 8.3. By Theorem 7,4 this is equal to the transcendence degree of $[(A/I_\sigma) \otimes K]$ over K. Now $[(A/I_\sigma) \otimes K]$ is isomorphic with $[\exp(t\sigma)(A \otimes K)] = L$, say. We have seen in proving Theorem 8.3 that

$$K(\exp(ta_1), \ldots, \exp(ta_q)) \subset L \subset K(t, \exp(ta_1), \ldots, \exp(ta_q)).$$

If σ^* is semisimple, then all the linear endomorphisms μ_i of the proof of Theorem 8.3 are 0, whence we see that $L = K(\exp(ta_1), \ldots, \exp(ta_q))$, whose transcendence degree over K is q. If σ^* is not semisimple, then not all of the μ_i's can be 0, whence we see that $L = K(t, \exp(ta_1), \ldots, \exp(ta_q))$, whose transcendence degree over K is $q + 1$. This proves Proposition 8.4.

PROPOSITION 8.5 *Let G be an affine algebraic group over an arbitrary field F. Let H and K be algebraic subgroups of G such that $H \subset K$. Then $\mathscr{L}(H) \subset \mathscr{L}(K)$, and we have $\mathscr{L}(H) = \mathscr{L}(K)$ only if $H_1 = K_1$.*

Proof: Let A be the algebra of polynomial functions on G, let I be the annihilator of H in A, and let J be the annihilator of K in A. Clearly, $J \subset I$. Here, $\mathscr{L}(H)$ and $\mathscr{L}(K)$ are viewed as Lie subalgebras of $\mathscr{L}(G)$. As such, $\mathscr{L}(H)$ is the annihilator of I in $\mathscr{L}(G)$, and $\mathscr{L}(K)$ is the annihilator of J. Hence it is clear that $\mathscr{L}(H) \subset \mathscr{L}(K)$. With these identifications, we have $\mathscr{L}(H) = \mathscr{L}(H_1)$ and $\mathscr{L}(K) = \mathscr{L}(K_1)$, as was shown in Section 7. Hence, what remains to be shown is that if H and K are connected algebraic subgroups of G such that $H \subset K$ and $\mathscr{L}(H) = \mathscr{L}(K)$, then $H = K$. Let α denote the canonical homomorphism $A/J \to A/I$. Choose $x_1, \ldots, x_n$ in A/J such that $(\alpha(x_1), \ldots, \alpha(x_n))$ is a transcendence basis of $[A/I]$ over F. Then $(x_1, \ldots, x_n)$ is algebraically independent over F. By Theorem 7.4, the assumption that $\mathscr{L}(H) = \mathscr{L}(K)$ means that $[A/I]$ and $[A/J]$ have the same transcendence degree over F. Hence $[A/J]$ is algebraic over $F(x, \ldots, x_n)$. Now suppose that there is a non-zero element y in A/J such that $\alpha(y) = 0$. There are elements f_i in $F[x_1, \ldots, x_n]$ such that $f_0 + \cdots + f_m y^m = 0$ and $f_m \neq 0$. Choosing these so that m is as small as possible, we obtain that also $f_0 \neq 0$. Applying α, we obtain $f_0(\alpha(x_1), \ldots, \alpha(x_n)) = 0$, which contradicts the algebraic independence of the $\alpha(x_i)$'s. Hence we conclude that α is injective, which means that $J = I$, so that $H = K$. This establishes Proposition 8.5.

The next theorem is the main result concerning the correspondence between algebraic subgroups and Lie subalgebras.

THEROEM 8.6 *Let G be an affine algebraic group over a field of characteristic 0. For every Lie subalgebra L of $\mathscr{L}(G)$, let G_L denote the* *intersection of the family of all algebraic subgroups of G whose Lie algebras contain L. Then G_L is a connected algebraic subgroup of G, and $L \subset \mathscr{L}(G_L)$. If H and K are connected algebraic subgroups of G, then $H \subset K$ if and only if $\mathscr{L}(H) \subset \mathscr{L}(K)$.*

Proof: Evidently, G_L is an algebraic subgroup of G. By Theorem 8.3, we have $G_\sigma \subset G_L$ for every element σ of L. Hence (Prop. 8.5) $\mathscr{L}(G_\sigma) \subset \mathscr{L}(G_L)$. From the last part of Theorem 8.3, we have therefore $\sigma \in \mathscr{L}(G_L)$. Thus we have shown that $L \subset \mathscr{L}(G_L)$. Since $\mathscr{L}(G_L) = \mathscr{L}((G_L)_1)$, this shows also that $G_L \subset (G_L)_1$, so that G_L is connected.

Let H be any connected algebraic subgroup of G. Since $G_{\mathscr{L}(H)} \subset H$, we have $\mathscr{L}(G_{\mathscr{L}(H)}) \subset \mathscr{L}(H)$. From the above, we have the reversed inclusion. Hence $\mathscr{L}(G_{\mathscr{L}(H)}) = \mathscr{L}(H)$. Since H and $G_{\mathscr{L}(H)}$ are connected, and since $G_{\mathscr{L}(H)} \subset H$, this implies, by Proposition 8.5, that $G_{\mathscr{L}(H)} = H$. Now if H and K are connected algebraic subgroups of G, and $\mathscr{L}(H) \subset \mathscr{L}(K)$, then evidently $G_{\mathscr{L}(H)} \subset G_{\mathscr{L}(K)}$, i. e., $H \subset K$. In view of the trivial part of Proposition 8.5, this establishes Theorem 8.6.

Over fields of non-zero characteristic, the correspondence between Lie subalgebras and algebraic subgroups is defective. We shall illustrate this by means of an example. Let F be any infinite field of non-zero characteristic p. Let G be the direct product of two copies of the multiplicative group of F, and let u and v be the two projections $G \to F^*$. Let A be the F-algebra of functions $G \to F$ that is generated by u, v and the reciprocals of u and v. One verifies immediately that A is a fully stable algebra of representative functions on G, and that (G, A) is the structure of an affine algebraic group over F. Moreover, it is easy to see that A is an integral domain, so that G is connected. Let H be the subgroup consisting of all elements of the form (a, a^{p-1}), and let K be the subgroup of all elements of the form (a, a^{-1}). One verifies directly that H and K are connected algebraic subgroups of G, their annihilating ideals I and J being generated by $v - u^{p-1}$ and $uv - 1$, respectively. Let δ be any element of $\mathscr{L}(G)$. Then

$$\delta(v - u^{p-1}) = \delta(v) - (p-1)\delta(u) = \delta(v) + \delta(u) = \delta(uv - 1)\,.$$

This shows that δ annihilates I if and only if it annihilates J, whence $\mathscr{L}(H) = \mathscr{L}(K)$, although we have $H \neq K$.

THEOREM 8.7 *Let $\rho : G \to H$ be a morphism of affine algebraic groups over a field of characteristic 0, and let K be the kernel of ρ. Then the kernel of the differential ρ° of ρ coincides with $\mathscr{L}(K)$.*

Proof: Clearly (for an arbitrary base field), $\mathscr{L}(K)$ is contained in the kernel of ρ°. Conversely, let σ be any element of the kernel of ρ°. Let f be a polynomial function on H, and let x be any element of G. Then we have

$$\begin{aligned}\sigma^*(f \circ \rho)(x) &= \sigma\big((f \circ \rho) \cdot x\big) \\ &= \sigma\big((f \cdot \rho(x)) \circ \rho\big) = \rho^\circ(\sigma)\big(f \cdot \rho(x)\big) = 0 .\end{aligned}$$

Thus $\sigma^*(f \circ \rho) = 0$ for every f in $\mathscr{A}(H)$. Hence the elements $f \circ \rho - c(f)$ of $\mathscr{A}(G)$ (where c stands for the co-unit of $\mathscr{A}(H)$) belong to the annihilator I_σ of G_σ, whence we conclude that $G_\sigma \subset K$. Using Theorem 8.3, we obtain from this that $\sigma \in \mathscr{L}(G_\sigma) \subset \mathscr{L}(K)$. This proves Theorem 8.7.

EXERCISES

1. Let G be an affine algebraic group over a field F of characteristic 0, and let H be an algebraic subgroup of G. Let P denote the field of power series in one variable t with coefficients in F (i. e., the field of fractions of $F[[t]]$). Show that an element σ of $\mathscr{L}(G)$ belongs to $\mathscr{L}(H)$ if and only if $\exp(t\sigma)$ belongs to the algebraic subgroup H^P of G^P.

2. Let G, F, P be as in Exercise 1, and let $\rho: G \to K$ be a morphism of affine algebraic groups. Let $\rho^P: G^P \to K^P$ be the morphism of affine algebraic groups over P that is obtained from ρ in the natural fashion. Show that, for every element σ of $\mathscr{L}(G)$, one has $\rho^P\big(\exp(t\sigma)\big) = \exp\big(t\rho^\circ(\sigma)\big)$ where ρ° is the differential of ρ.

9. POLYNOMIAL REPRESENTATIONS

Let V be a finite-dimensional vector space over a field F. We denote the F-algebra of all F-linear endomorphisms of V by $\mathscr{E}(V)$, and the group of all F-linear automorphisms by $\mathscr{G}(V)$. The restrictions to $\mathscr{G}(V)$ of the elements of the dual space $\mathscr{E}(V)^\circ$, together with the reciprocal of the determinant function on $\mathscr{G}(V)$, generate a fully stable F-algebra P of representative functions on $\mathscr{G}(V)$, and it is clear from elementary linear algebra that every F-algebra homomorphism $P \to F$ is the evaluation at some element of $\mathscr{G}(V)$. Thus $(\mathscr{G}(V), P)$ is the structure of an affine algebraic group over F. If F is an infinite field, then the image in P of an F-basis of $\mathscr{E}(V)^\circ$ is algebraically free over F, whence it is clear that P is then an integral domain, i.e., that $\mathscr{G}(V)$ is connected. It is always true that the restriction map $\mathscr{E}(V)^\circ \to P$ is injective, because $\mathscr{G}(V)$ always spans $\mathscr{E}(V)$ over F (see D. Zelinsky, Proc. Am. Math. Soc., vol. 5 (1954), pp. 627–630). Accordingly, *we shall identify* $\mathscr{E}(V)^\circ$ *with its image in* P.

Let (G, A) be the structure of an affine algebraic group over F. By a *polynomial representation* of G on V we mean a morphism of affine algebraic groups $G \to \mathscr{G}(V)$. Equivalently, a polynomial representation of G on V is the structure of a G-module on V such that the associated space $\mathscr{S}(V)$ of representative functions on G lies in A. Let ρ be such a polynomial representation, and consider the corresponding A-comodule structure $\rho^* : V \to V \otimes A$, as defined in Section 2. For every v in V, we have $\rho^*(v) = \sum_i v_i \otimes f_i$, where each f_i belongs to $\mathscr{S}(V)$, and it follows immediately from the definition of ρ^* that we then have

$$\rho^*(\rho(x)(v)) = \sum_i v_i \otimes x \cdot f_i$$

for every x in G. This shows that ρ^*, which we know to be injective, is a G-module isomorphism of V onto a submodule of a direct sum of copies of $\mathcal{S}(V)$. Thus *every polynomial representation G-module is isomorphic with a G-submodule of a direct sum of copies of a submodule of the G-module A of polynomial functions on G.*

There are two immediate important consequences of this, as follows. We say that ρ is a *unipotent representation* if the algebra generated in $\mathcal{E}(V)$ by the endomorphisms $\rho(x) - i_V$, with x ranging over G, is nilpotent.

(1) *If the representation of G by left translations on A is locally unipotent* (i.e., *unipotent on every finite-dimensional left stable subspace of A*), *then every polynomial representation of G is unipotent,*

(2) *If A is semisimple as a G-module, then every polynomial representation of G is semisimple.*

Next, we examine the differentials of polynomial representations. For this purpose, we develop an explicit description of the Lie algebra of $\mathcal{G}(V)$, where V is a finite-dimensional vector space over the field F. Let δ be an element of this Lie algebra. The restriction of δ to $\mathcal{E}(V)^\circ \subset P$ is an F-linear functional on $\mathcal{E}(V)^\circ$, and is therefore the evaluation at a uniquely defined element δ' of $\mathcal{E}(V)$. Since the only differentiation of P that annihilates $\mathcal{E}(V)^\circ$ is the 0-map, it is clear that the map $\delta \to \delta'$ is an injective linear map of the Lie algebra of $\mathcal{G}(V)$ into $\mathcal{E}(V)$. Let us regard $\mathcal{E}(V)$ as a Lie algebra, with the Lie composition $(e_1, e_2) \to e_1e_2 - e_2e_1$. We shall show that the map $\delta \to \delta'$ is then a homomorphism of Lie algebras. Evidently, $\mathcal{E}(V)^\circ$ is bistable as a subspace of P. Hence, if γ is the comultiplication of P, and if f is an element of $\mathcal{E}(V)^\circ$, we have $\gamma(f) = \sum_i g_i \otimes f_i$, where the g_i's and the h_i's belong to $\mathcal{E}(V)^\circ$. Now let δ_1 and δ_2 be elements of $\mathcal{L}(\mathcal{G}(V))$. Then we have

$$\begin{aligned} f([\delta_1, \delta_2]') &= [\delta_1, \delta_2](f) = (\delta_1 \otimes \delta_2 - \delta_2 \otimes \delta_1)(\gamma(f)) \\ &= \sum_i \big(\delta_1(g_i)\delta_2(h_i) - \delta_2(g_i)\delta_1(h_i)\big) \\ &= \sum_i \big(g_i(\delta_1')h_i(\delta_2') - g_i(\delta_2')h_i(\delta_1')\big). \end{aligned}$$

If x and y are elements of $\mathcal{G}(V)$, we have

$$f(xy) = (x \otimes y)(\gamma(f)) = \sum_i g_i(x)h_i(y).$$

Since the elements of $\mathcal{G}(V)$ span $\mathcal{E}(V)$ over F, it follows that this last equality holds also when x and y are arbitrary elements of $\mathcal{E}(V)$. Hence the above gives $f([\delta_1, \delta_2]') = f(\delta_1'\delta_2' - \delta_2'\delta_1')$. Since this holds for every f in $\mathscr{E}(V)^\circ$, it follows that $[\delta_1, \delta_2]' = \delta_1'\delta_2' - \delta_2'\delta_1'$. Thus *the map $\delta \to \delta'$ defined above is an injective Lie algebra homomorphism $\mathscr{L}(\mathscr{G}(V)) \to \mathscr{E}(V)$.* Accordingly, we shall usually identify $\mathscr{L}(\mathscr{G}(V))$ with its image in $\mathscr{E}(V)$. *If F is infinite,* one sees readily from what we said at the beginning of this section about the image of $\mathscr{E}(V)^\circ$ in P that *the map $\delta \to \delta'$ is bijective, and thus is an isomorphism of Lie algebras, by means of which we may identify $\mathscr{L}(\mathscr{G}(V))$ with $\mathscr{E}(V)$.*

If $\rho : G \to \mathscr{G}(V)$ is a polynomial representation of the affine algebraic group G, then the differential ρ°, viewed as a Lie algebra homomorphism $\mathscr{L}(G) \to \mathscr{E}(V)$, can be expressed quite simply in terms of the comodule structure $\rho^* : V \to V \otimes A$. In fact,

$$\rho^\circ(\tau) = (i_V \otimes \tau) \circ \rho^*$$

for every element τ of $\mathscr{L}(G)$. In order to verify this, let μ be an element of V°, and let v be an element of V. Define the element μ/v of $\mathscr{E}(V)^\circ$ by $(\mu/v)(e) = \mu(e(v))$ for every e in $\mathscr{E}(V)$. Then $(\mu/v) \circ \rho$ is an element of A, and our identification of $\mathscr{L}(\mathscr{G}(V))$ with its image in $\mathscr{E}(V)$ allows us to write $\tau((\mu/v) \circ \rho) = (\mu/v)(\rho^\circ(\tau))$. From the definition of ρ^*, we have $(\mu/v) \circ \rho = (\mu \otimes i_A)(\rho^*(v))$. Hence the last equality becomes

$$\mu(\rho^\circ(\tau)(v)) = (\mu \otimes \tau)(\rho^*(v)) .$$

Since this holds for every μ in V°, we have therefore

$$\rho^\circ(\tau)(v) = (i_V \otimes \tau)(\rho^*(v)) , \quad \text{q.e.d.}$$

From this expression for $\rho^\circ(\tau)$, we obtain the following useful relation between $\rho^\circ(\tau)$ and the proper derivation τ^* of A that corresponds to τ:

$$\rho^* \circ \rho^\circ(\tau) = (i_V \otimes \tau^*) \circ \rho^* .$$

The verification of this is straightforward, using the formula

$$(\rho^* \otimes i_A) \circ \rho^* = (i_V \otimes \gamma) \circ \rho^* ,$$

where γ is the comultiplication of A.

THEOREM 9.1. *Let (G, A) be the structure of a connected affine algebraic group over a field F, and let $\rho : G \to \mathscr{G}(V)$ be a polynomial representation of G. Then every product of elements of $\rho^\circ(\mathscr{L}(G))$ is an F-linear combination of endomorphisms of the form $\rho(x) - i_V$ with x in G. If F is of characteristic 0, then every endomorphism $\rho(x) - i_V$ is an F-linear combination of products of elements of $\rho^\circ(\mathscr{L}(G))$.*

Proof: Let $\sigma_1, \ldots, \sigma_n$ be elements of $\mathscr{L}(G)$, and let v be an element of V. From the last identity derived above, we obtain (by iteration)

$$\rho^*\big((\rho^\circ(\sigma_1) \cdots \rho^\circ(\sigma_n))(v)\big) = (i_V \otimes \sigma_1^* \cdots \sigma_n^*)\big(\rho^*(v)\big).$$

Let c denote the co-unit of A, and let μ be an element of V°. Apply the linear functional $\mu \otimes c$ to the above, nothing that $(i_V \otimes c) \circ \rho^* = i_V$, and that $(\mu \otimes i_A)\big(\rho^*(v)\big) = (\mu/v) \circ \rho$. This yields

$$(\mu/v)\big(\rho^\circ(\sigma_1) \cdots \rho^\circ(\sigma_n)\big) = c\big((\sigma_1^* \cdots \sigma_n^*)((\mu/v) \circ \rho)\big).$$

Since the elements of the form μ/v span $\mathscr{E}(V)^\circ$ over F, it follows that

$$\alpha\big(\rho^\circ(\sigma_1) \cdots \rho^\circ(\sigma_n)\big) = c\big((\sigma_1^* \cdots \sigma_n^*)(\alpha \circ \rho)\big)$$

for every element α of $\mathscr{E}(V)^\circ$. Now suppose that α annihilates every endomorphism of the form $\rho(x) - i_V$. Then $\alpha \circ \rho$ is constant on G, and the above shows that α therefore annihilates every product of elements of $\rho^\circ\big(\mathscr{L}(G)\big)$. Hence we conclude that every such product must belong to the F-space spanned by the endomorphisms of the form $\rho(x) - i_V$.

Now let us assume that F is of characteristic 0. Let β be an element of $\mathscr{E}(V)^\circ$ that annihilates every power of $\rho^\circ(\sigma)$, where σ is an arbitrarily fixed element of $\mathscr{L}(G)$. Then we see at once from the above identity that $c\big(\sigma^{*n}(\beta \circ \rho)\big) = 0$ for every positive exponent n. Hence the element $\beta \circ \rho - c(\beta \circ \rho)$ of A belongs to the annihilator I_σ of the subgroup G_σ of G, as discussed in Section 8. This means that β annihilates every endomorphism of the form $\rho(x) - i_V$, with x in G_σ. It follows that every such endomorphism is equal to an F-linear combination of powers (with positive exponents) of $\rho^\circ(\sigma)$. Let us now consider an element α of $\mathscr{E}(V)^\circ$ that annihilates every product of elements of $\rho^\circ\big(\mathscr{L}(G)\big)$. Let $\sigma_1 \ldots, \sigma_n$ be elements of $\mathscr{L}(G)$, and let x_i be an element of G_{σ_i}. Then the endomorphism $\rho(x_1 \cdots x_n) - i_V$ can be written as an integral linear combination of products of the endomorphisms $\rho(x_i) - i_V$, and therefore, by what we have shown above, is an F-linear combination of products of elements of $\rho^\circ\big(\mathscr{L}(G)\big)$. Hence α annihilates every endomorphism of the form $\rho(x) - i_V$, with x in the group generated by all the groups G_σ. Therefore, α must also annihilate $\rho(x) - i_V$ whenever x belongs to the algebraic hull, H say, of the group generated by the G_σ's. We see immediately from Theorem 8.3 that $\mathscr{L}(H)$ must coincide with $\mathscr{L}(G)$. Therefore, by Proposition 8.5, we must have $H = G$. Thus α annihilates every endomorphism of the form $\rho(x) - i_V$. It follows that every such endomorphism must be an F-linear combination of products of elements of $\rho^\circ\big(\mathscr{L}(G)\big)$, and this completes the proof of Theorem 9.1.

The following is an immediate corollary. In stating it, we suppress the explicit reference to the polynomial representation.

COROLLARY 9.2. *Let G be a connected affine algebraic group over a field F, and let V be a polynomial representation G-module. Then the G-fixed part of V is annihilated by $\mathscr{L}(G)$. If F is of characteristic* 0, *then the G-fixed part of V coincides with the $\mathscr{L}(G)$-annihilated part.*

In Section 2, we exhibited the comodule structure that corresponds to the usual tensor product of group modules. We use this to compute the differential of a tensor product representation. Let $\rho : G \to \mathscr{G}(U)$ and $\sigma : G \to \mathscr{G}(V)$ be two polynomial representations of the affine algebraic group G over the field F. Write A for $\mathscr{A}(G)$. Then the tensor product comodule structure is the map

$$(i_U \otimes i_V \otimes \mu) \circ (i_U \otimes s_{A,V} \otimes i_A) \circ (\rho^* \otimes \sigma^*)$$

of $U \otimes V$ into $U \otimes V \otimes A$, where μ is the multiplication of A, and $s_{A,V}$ is the switching map $A \otimes V \to V \otimes A$. Let π denote this comodule structure, and let δ be any element of $\mathscr{L}(G)$. We have $\pi = (\rho \otimes \sigma)^*$, where $\rho \otimes \sigma$ stands for the usual tensor product representation of G on $U \otimes V$. From the first of the two identities we established just before stating Theorem 9.1, we have therefore

$$(\rho \otimes \sigma)^\circ(\delta) = (i_U \otimes i_V \otimes \delta) \circ \pi \, .$$

On substituting the above expression for π, this becomes

$$(\rho \otimes \sigma)^\circ(\delta) = (i_U \otimes i_V \otimes \delta \circ \mu) \circ (i_U \otimes s_{A,V} \otimes i_A) \circ (\rho^* \otimes \sigma^*) \, .$$

Now $\delta \circ \mu = \delta \otimes c + c \otimes \delta$, so that the expression on the right is equal to

$$\begin{aligned}(i_U \otimes \delta \otimes i_V \otimes c + i_U \otimes c \otimes i_V \otimes \delta) \circ (\rho^* \otimes \sigma^*)\\ = (i_U \otimes \delta) \circ \rho^* \otimes i_V + i_U \otimes (i_V \otimes \delta) \circ \sigma^*\\ = \rho^\circ(\delta) \otimes i_V + i_U \otimes \sigma^\circ(\delta) \, .\end{aligned}$$

Thus *the differential of the tensor product representation $\rho \otimes \sigma$ of G on $U \otimes V$ is given by*

$$(\rho \otimes \sigma)^\circ(\delta) = \rho^\circ(\delta) \otimes i_V + i_U \otimes \sigma^\circ(\delta) \, .$$

Next, we discuss the adjoint representation of an affine algebraic group G on its Lie algebra. For every element x of G, let τ_x denote the conjugation effected by x on G; $\tau_x(y) = xyx^{-1}$. Clearly, τ_x is an automorphism of affine algebraic groups, and we can form its differential $\tau_x{}^\circ : \mathscr{L}(G) \to$

$\mathscr{L}(G)$. The map $x \to \tau_x{}^\circ$ is a polynomial representation of G on $\mathscr{L}(G)$. This is called the *adjoint representation* of G, and we shall denote it by α. Thus $\alpha(x) = \tau_x{}^\circ$ for every element x of G. Observe that the Lie algebra automorphisms of $\mathscr{L}(G)$ constitute an algebraic subgroup $\mathscr{A}ut(\mathscr{L}(G))$ of $\mathscr{G}(\mathscr{L}(G))$, and that α may also be regarded as a morphism $G \to \mathscr{A}ut(\mathscr{L}(G))$.

The adjoint representation is more transparently realized within the algebra of the proper endomorphisms of $\mathscr{A}(G)$. Paralleling the notation we use for Lie algebra elements, let x^* denote the left translation effected by the element x on $\mathscr{A}(G)$. Let σ be an element of $\mathscr{L}(G)$, and let a be an element of $\mathscr{A}(G)$. From the definition of the adjoint representation α, we have

$$\alpha(x)(\sigma)(a) = \sigma(x^{-1} \cdot a \cdot x) = \sigma\big(x^{*-1}(a) \cdot x\big) = \sigma^* x^{*-1}(a)(x)$$
$$= c\big(x^*\sigma^*x^{*-1}(a)\big),$$

whence

$$\alpha(x)(\sigma)^* = x^*\sigma^*x^{*-1}.$$

Evidently, the center of G lies in the kernel of α. If the base field is of characteristic 0, we have the following precise result.

THEOREM 9.3. *Let G be a connected affine algebraic group over a field of characteristic* 0. *Then the kernel of the adjoint representation coincides with the center of G.*

Proof: Let V be a finite-dimensional left stable subspace of $\mathscr{A}(G)$, and consider the representation of G on V that sends every element x of G onto the restriction to V of x^*. If ρ denotes this representation, then the corresponding comodule structure $\rho^* : V \to V \otimes A$ is the restriction to V of the comultiplication γ of $\mathscr{A}(G)$. Hence we see from our standard expression of the differential in terms of the comodule structure that $\rho^\circ(\sigma) = (i_V \otimes \sigma) \circ \gamma = \sigma_V{}^*$ for every element σ of $\mathscr{L}(G)$. Now let x be any element of the kernel of the adjoint representation. Then x^* commutes with every σ^*. From Theorem 9.1, it follows that the restriction of x^* to V commutes with the restriction to V of every y^*, with y in G. Letting V range over the family of all finite-dimensional left stable subspaces of $\mathscr{A}(G)$, we conclude that x^* commutes with every y^*. Clearly, this implies that x lies in the center of G, so that Theorem 9.3 is proved.

The following example shows that Theorem 9.3 fails when the base field is of non-zero characteristic. Let F be an infinite field of non-zero characteristic p. We define a group G consisting of the pairs (r, s) with r and s in F, and with $s \neq 0$. The product of two such pairs is defined by

$$(r_1, s_1)(r_2, s_2) = (r_1 + s_1{}^p r_2, s_1 s_2) .$$

Define F-valued functions u and v on this group by

$$u(r, s) = r, \qquad v(r, s) = s .$$

Let A denote the F-algebra of functions on G that is generated by u, v, and the reciprocal v^{-1} of v. It is easily verified that (G, A) is the structure of an affine algebraic group over F, and that the comultiplication γ of A is given by

$$\gamma(u) = u \otimes 1 + v^p \otimes u , \qquad \gamma(v) = v \otimes v .$$

We claim that every element of the form $(r, 1)$ is in the kernel of the adjoint representation of G. In order to prove this, it suffices to show that $(r, 1)^*$ commutes with every σ^*, with σ in $\mathscr{L}(G)$. We have

$$c \circ \big((r, 1)^* \sigma^*\big) = \big((r, 1) \otimes \sigma\big) \circ \gamma ,$$

and we see immediately from the above description of γ that the function on the right hand side coincides with σ. Using the fact that the differentiation σ annihilates the p-th power v^p, we see similarly that $c \circ \big(\sigma^*(r, 1)^*\big)$ also coincides with σ. Since the composition with the co-unit c is injective on the space of the proper endomorphisms of A, this proves our above claim. On the other hand, it is clear that $(r, 1)$ belongs to the center of G only if $r = 0$.

Next, let us compute the differential α° of the adjoint representation α of an arbitrary affine algebraic group G. Write A for $\mathscr{A}(G)$, and let $\alpha^* : \mathscr{L}(G) \to \mathscr{L}(G) \otimes A$ be the comodule structure defined by α. Then, if σ and τ are elements of $\mathscr{L}(G)$, we have

$$\alpha^\circ(\sigma)(\tau) = (i_{\mathscr{L}(G)} \otimes \sigma)\big(\alpha^*(\tau)\big) .$$

In order to obtain a convenient expression for $\alpha^*(\tau)$, we view the elements of $\mathscr{L}(G) \otimes A$ as linear maps $A \to A$ such that $(\rho \otimes a)(b) = \rho(b)a$ for every element ρ of $\mathscr{L}(G)$ and every element a of A. Then we have, for every x in G,

$$\alpha^*(\tau)(a)(x) = \alpha(x)(\tau)(a) = \tau(x^{-1} \cdot a \cdot x) ,$$

as is seen directly from the definitions. The map sending each element a of A onto $x^{-1} \cdot a \cdot x$ is $(i_A \otimes x \circ \eta) \circ \gamma \circ (x \otimes i_A) \circ \gamma$, where γ is the comultiplication, and η is the antipode of A. Using the co-associativity of γ, we can rewrite this map as $((x \otimes \cdot i_A) \circ \gamma \otimes x \circ \eta) \circ \gamma$. Hence the map sending each a onto $\tau(x^{-1} \cdot a \cdot x)$ is

$$((x \otimes \tau) \circ \gamma \otimes x \circ \eta) \circ \gamma = (x \otimes x) \circ (\tau^* \otimes \eta) \circ \gamma = x \circ \mu \circ (\tau^* \otimes \eta) \circ \gamma,$$

where μ is the multiplication of A. Thus, as a map $A \to A$, $\alpha^*(\tau)$ is equal to $\mu \circ (\tau^* \otimes \eta) \circ \gamma$. Substituting this in our above expression for $\alpha^\circ(\sigma)(\tau)$, we obtain

$$\alpha^\circ(\sigma)(\tau) = \sigma \circ \mu \circ (\tau^* \otimes \eta) \circ \gamma .$$

Writing $\sigma \circ \mu = \sigma \otimes c + c \otimes \sigma$, we obtain from this that

$$\alpha^\circ(\sigma)(\tau) = (\sigma \circ \tau^* \otimes c \circ \eta + \tau \otimes \sigma \circ \eta) \circ \gamma .$$

Now $c \circ \eta = c$, and $(\sigma \circ \tau^* \otimes c) \circ \gamma = \sigma \circ \tau^* = (\sigma \otimes \tau) \circ \gamma$. Thus

$$\alpha^\circ(\sigma)(\tau) = (\sigma \otimes \tau + \tau \otimes \sigma \circ \eta) \circ \gamma .$$

Now recall from Section 2 that $\mu \circ (\eta \otimes i_A) \circ \gamma = c$, whence

$$\sigma \circ \mu \circ (\eta \otimes i_A) \circ \gamma = 0 , \quad \text{i.e.,}$$

$$(\sigma \circ \eta \otimes c + c \circ \eta \otimes \sigma) \circ \gamma = 0 , \quad \text{i.e.,} \quad \sigma \circ \eta + \sigma = 0 .$$

Hence the above becomes

$$\alpha^\circ(\sigma)(\tau) = (\sigma \otimes \tau - \tau \otimes \sigma) \circ \gamma = [\sigma, \tau] .$$

For any Lie algebra L, the map that sends each element σ of L onto the derivation $\tau \to [\sigma, \tau]$ of L is a Lie algebra homomorphism of L into the Lie algebra of its derivations, and this is called the *adjoint representation* of L. Our result may therefore be expressed by saying that *the differential of the adjoint representation of an affine algebraic group is the adjoint representation of its Lie algebra.*

The next theorem is a simple application of this result. A Lie algebra L is called *abelian* if $[L, L] = (0)$.

THEOREM 9.4. *Let G be an affine algebraic group over a field F. If G is abelian, so is $\mathscr{L}(G)$. The converse holds whenever G is connected and F is of characteristc* 0.

Proof: If G is abelian, then the adjoint representation of G is trivial, whence its differential is trivial. By the last result above, this means precisely that $\mathscr{L}(G)$ is abelian.

Now suppose that G is connected, that F is of characteristic 0, and that $\mathscr{L}(G)$ is abelian. From the last assumption, we have that the algebra of endomorphisms of $\mathscr{A}(G)$ that is generated by the elements

σ^*, with σ in $\mathscr{L}(G)$, is commutative. By Theorem 9.1 (applied to the representations of G by left translations on the finite-dimensional left stable subspaces of $\mathscr{A}(G)$), this implies that the left translations effected by the elements of G on $\mathscr{A}(G)$ commute with each other, so that G is abelian. This proves Theorem 9.4

THEOREM 9.5. *Let G be a connected affine algebraic group over the field F, and let K be a connected algebraic subgroup of G. If K is normal in G, then $\mathscr{L}(K)$ is an ideal of $\mathscr{L}(G)$. Conversely, if F is of characteristic* 0, *and if $\mathscr{L}(K)$ is an ideal of $\mathscr{L}(G)$, then K is normal in G.*

Proof: Suppose that K is normal in G, let x be an element of G, let σ be an element of $\mathscr{L}(K)$, and let I denote the annihilator of K in $\mathscr{A}(G)$. Evidently, $x^*\sigma^*x^{*-1}$ stabilizes I, so that it belongs to $\mathscr{L}(K)^*$. We know that, if α denotes the adjoint representation of G, then

$$\alpha(x)(\sigma)^* = x^*\sigma^*x^{*-1} .$$

Hence we conclude that $\mathscr{L}(K)$ is $\alpha(G)$-stable. By Theorem 9.1, this implies that $\mathscr{L}(K)$ is also $\alpha^\circ(\mathscr{L}(G))$-stable, i.e., that $[\mathscr{L}(G), \mathscr{L}(K)] \subset \mathscr{L}(K)$. Thus $\mathscr{L}(K)$ is an ideal in $\mathscr{L}(G)$ whenever K is normal in G.

Now suppose that F is of characteristic 0, and that $\mathscr{L}(K)$ is an ideal in $\mathscr{L}(G)$. Then we have from Theorem 9.1 that $\mathscr{L}(K)$ is $\alpha(G)$-stable, so that $x^*\sigma^*x^{*-1}$ stabilizes I for every x in G, and every σ in $\mathscr{L}(K)$. Thus, for every x in G, the ideal $x^{-1} \cdot I \cdot x$ of $\mathscr{A}(G)$ is stable under the action of $\mathscr{L}(K)$ by proper derivations on $\mathscr{A}(G)$. Applying Theorem 9.1 to the polynomial representations of K on the finite-dimensional left stable subspaces of $\mathscr{A}(G)$, with K acting by left translations, we conclude from this that $x^{-1} \cdot I \cdot x$ is stable also under the left translation action of K. Hence xKx^{-1} belongs to the annihilator of I in G, i.e., $xKx^{-1} \subset K$. Thus K is normal in G, and Theorem 9.5 is proved.

EXERCISES

1. Let G be an affine algebraic group, and let H be an algebraic subgroup of G. A polynomial representation H-module is said to be *extendible* if it is contained as an H-submodule in a polynomial representation G-module. From the discussion at the beginning of this section, show that every polynomial representation H-module is isomorphic with a factor module U/V, where U is an extendible polynomial representation H-module. As a consequence, if $\mathscr{A}(H)$ is semisimple as an H-module, then every polynomial representation H-module is extendible.

2. For any vector space T, let $E^n(T)$ denote the homogeneous component of degree n of the exterior algebra built over T. Note that any polynomial representation on T yields one on $E^n(T)$ in a natural way, via the tensor product of polynomial representations. Retaining the notation of Exercise 1, let n be the dimension of V. By considering the H-submodule $UE^n(V)$ of $E^{n+1}(U)$, prove the following: if the dual of every extendible 1-dimensional polynomial representation H-module is extendible, then every polynomial representation H-module is extendible. As a consequence, if the representation of H on $\mathscr{A}(H)$ is locally unipotent, then every polynomial representation H-module is extendible.

3. Let E be a finite-dimensional associative algebra with identity element over an infinite field, and let $U(E)$ be the group of units of E, with the structure of an affine algebraic group as in Exercise 1. of Section 3. Generalizing the result of this section concerning the Lie algebra of $\mathscr{G}(V)$, show that the Lie algebra of $U(E)$ may be identified with E, when E is regarded as a Lie algebra with the Lie composition $[e_1, e_2] = e_1e_2 - e_2e_1$.

10. UNIPOTENT GROUPS

Let (G, A) be the structure of an affine algebraic group over an arbitrary field F. A subgroup T of G is called *unipotent* if, for every finite-dimensional left stable F-subspace V of A, the representation of T by left translations on V is unipotent (in the sense defined at the beginning of Section 9). We express this property also by saying that T is locally unipotent on A, or simply that T is A-unipotent. Correspondingly, a Lie subalgebra L of $\mathscr{L}(G)$ is said to be *A-nilpotent* if the representation of L by proper derivations of A is nilpotent on every finite-dimensional left stable subspace of A, i. e., if for every such subspace V, there is a positive integer n such that the product of any n proper derivations σ^*, with σ in L, annihilates V. We use the same terminology for elements of $\mathscr{L}(G)$, or of G.

Let T be a unipotent subgroup of G, and let V be a finite-dimensional left stable F-subspace of A. Let $(0) = V_n \subset \cdots \subset V_0 = V$ be a composition series of V as a T-module. The unipotency of T evidently implies that the representation of T on each V_i/V_{i+1} is trivial. This is expressible by the condition that T annihilate certain elements of A. It follows that, if $[T]$ denotes the algebraic hull of T in G, then the V_i's also constitute a composition series for V as a $[T]$-module, and the representation of $[T]$ on each V_i/V_{i+1} is trivial. In particular, this shows that $[T]$ is a unipotent subgroup of G. Moreover, we see immediately from Theorem 9.1 that the V_i's are stable under the action of $\mathscr{L}([T])$, and that the induced representation of $\mathscr{L}([T])$ on each V_i/V_{i+1} is trivial, i. e., that the proper derivations σ^*, with σ in $\mathscr{L}([T])$, map each V_i into V_{i+1}. Hence it is clear that $\mathscr{L}([T])$ is an A-nilpotent Lie subalgebra of $\mathscr{L}(G)$. Thus *the algebraic hull of a unipotent subgroup of G is unipotent, and its Lie algebra is A-nilpotent.* In the case where F is of characteristic 0, we have the following much stronger result.

10.

THEOREM 10.1 *Let (G, A) be the structure of an affine algebraic group over a field F of characteristic* 0. *Let T be a unipotent algebraic subgroup of G. Then T is connected, and the map $\tau \to c \circ \exp(\tau^*)$ is an isomorphism of affine algebraic varieties $\mathscr{L}(T) \to T$, where the algebraic variety structure of $\mathscr{L}(T)$ is the canonical one, obtained from its F-space structure. In this way, the family of all unipotent algebraic subgroups of G is in bijective correspondence with the family of all A-nilpotent Lie subalgebras of $\mathscr{L}(G)$.*

Proof: Let T_1 denote the connected component of the neutral element in T. Denoting the algebra of polynomial functions on T by B, consider the representation of T by left translations on the T_1-fixed part B^{T_1} of B. Since T is unipotent, this is a locally unipotent representation of T. On the other hand, this representation factors through the finite group T/T_1. Because F is of characteristic 0, this implies that our representation of T is a semisimple representation. Evidently, a representation that is both semisimple and locally unipotent must be trivial. Hence we conclude that $B^{T_1} = B^T$. By Theorem 6.2, this implies that $T_1 = T$, so that T is connected.

Let e be any proper derivation of A that is locally nilpotent. Then the exponential series $\sum_{n=0}^{\infty} (n!)^{-1} e^n$ is evidently meaningful as a proper linear endomorphism $\exp(e)$ of A. From the fact that e is a derivation, it follows, via familiar formal considerations, that $\exp(e)$ is an F-algebra endomorphism of A. Hence $c \circ \exp(e)$, where c is the co-unit of A, is an element of G. Evidently, $\exp(e)$ is locally unipotent on A.

Conversely, let u be any locally unipotent proper F-algebra endomorphism of A. Then we can define its *logarithm* as a proper linear endomorphism of A by putting

$$\log(u) = \log\big(1 - (1 - u)\big) = -\sum_{n=1}^{\infty} n^{-1}(1 - u)^n .$$

It is a familiar formal matter to verify that $\exp\big(\log(u)\big) = u$, and $\log\big(\exp(e)\big) = e$. From the fact that u is an F-algebra homomorphism, we show that $\log(u)$ is an F-algebra derivation, as follows. Let t be an auxiliary variable, and work in the polynomial ring $A[t]$. Given elements a and b in A, there is a positive exponent m such that $\log(u)^{m+1}$ annihilates a, b, and ab. Then, if f is any one of these elements, we have

$$\exp\big(t \log(u)\big)(f) = \sum_{n=0}^{m} (n!)^{-1} t^n \log(u)^n(f) .$$

Hence the power series

$$\exp\bigl(t \log (u)\bigr)(ab) - \exp\bigl(t \log (u)\bigr)(a) \exp\bigl(t \log (u)\bigr)(b)$$

is actually a polynomial $p(t)$ in $A[t]$. For every non-negative integer k, we have $p(k) = u^k(ab) - u^k(a)u^k(b) = 0$. It follows that the polynomial $p(t)$ must have all its coefficients equal to 0. In particular, equating the coefficient of t to 0, we obtain

$$\log(u)(ab) = \log(u)(a)b + a \log(u)(b) .$$

Thus $\log(u)$ is indeed a derivation. Therefore, $c \circ \log(u)$ is an element of $\mathscr{L}(G)$.

Now let L be any A-nilpotent Lie subalgebra of $\mathscr{L}(G)$. Clearly, L is then a nilpotent Lie algebra, in the sense of ordinary Lie algebra theory. This allows us to utilize the *Campbell-Hausdorff formula*, as follows. Let Q denote the field of rational numbers, and let $Q\bigl[[x, y]\bigr]$ be the Q-algebra of integral formal power series in the two freely non-commuting variables x and y, with coefficients in Q. The Campbell-Hausdorff formula gives an element $s(x, y)$ of $Q\bigl[[x, y]\bigr]$ that can be written as a series whose terms are multiple commutators formed from x and y (the homogeneous component of degree 1 being $x + y$), such that

$$\exp(x) \exp(y) = \exp\bigl(s(x, y)\bigr) .$$

Now let σ and τ be elements of L. Since L is nilpotent, the formal substitution, in $s(x, y)$ of σ for x and τ for y, yields a finite linear combination of multiple commutators formed with σ and τ (including σ and τ themselves), and thus defines an element $s(\sigma, \tau)$ of L. If we write $\exp(\sigma)$ for $c \circ \exp(\sigma^*)$, etc., we have then

$$\exp(\sigma) \exp(\tau) = \exp\bigl(s(\sigma, \tau)\bigr) .$$

Thus the elements $\exp(\sigma)$, with σ in L, constitute a subgroup $\exp(L)$ of G. Clearly, $\exp(L)$ is a unipotent subgroup of G, and an element x of G that is locally unipotent on A (as translation operator) belongs to $\exp(L)$ if and only if $c \circ \log(x^*)$ belongs to L, where x^* denotes the left translation of A corresponding to x. We shall abbreviate $c \circ \log(x^*)$ by $\log(x)$. For an element x of the algebraic hull of $\exp(L)$, the test for the relation $x \in \exp(L)$ can be made by considering the restrictions of x^* to the finite-dimensional left stable F-subspaces of A. Since, for each of these, $\log(x^*)$ reduces to a polynomial in x^*, it follows that $\exp(L)$ is an algebraic subset of its algebraic hull, and therefore coincides with its algebraic hull. Thus $\exp(L)$ is a unipotent algebraic subgroup of G.

Now let I denote the annihilator of $\exp(L)$ in A, and let σ be an element of L. Then $\exp(\sigma)^*$ stabilizes I, i.e., $\exp(\sigma^*)$ stabilizes I. It follows that $\log(\exp(\sigma^*))$ stabilizes I, i. e., that σ^* stabilizes I, whence σ belongs to $\mathscr{L}(\exp(L))$. Thus we have $L \subset \mathscr{L}(\exp(L))$. Conversely, suppose that σ is an element of $\mathscr{L}(\exp(L))$. Then σ^* stabilizes I, whence also $\exp(\sigma^*)$ stabilizes I, which shows that $\exp(\sigma)$ belongs to $\exp(L)$. Applying log, we obtain from this that σ belongs to L. Hence we have $L = \mathscr{L}(\exp(L))$. It is clear from the above that log is a polynomial map of $\exp(L)$ onto L. Similarly, exp is a polynomial map of L onto $\exp(L)$. Since log and exp are mutually inverse, this shows that $\exp : L \to \exp(L)$ is an isomorphism of affine algebraic varieties, so that Theorem 10.1 is proved.

THEOREM 10.2. *Let (G, A) be the structure of a unipotent affine algebraic group over a field of characteristic 0. Let $\rho : G \to H$ be a morphism of affine algebraic groups. Then $\rho(G)$ is a unipotent algebraic subgroup of H. If ρ° is the differential of ρ, then $\rho \circ \exp = \exp \circ \rho^\circ$.*

Proof: For every finite-dimensional left stable subspace V of $\mathscr{A}(H)$, the morphism ρ defines a polynomial representation of G on V in the evident way. It is clear from our definitions that Theorem 10.2 will follow immediately as soon as we have established it in the special case where ρ is the polynomial representation $G \to \mathscr{G}(V)$, which we know to be unipotent, from Section 9. In this case, let $\rho^* : V \to V \otimes A$ denote the comodule structure defined by ρ. Then we have, for σ in $\mathscr{L}(G)$,

$$\rho(\exp(\sigma)) = (i_V \otimes \exp(\sigma)) \circ \rho^* .$$

Moreover,

$$i_V \otimes \exp(\sigma) = i_V \otimes c \circ \exp(\sigma^*) = (i_V \otimes c) \circ \exp(i_V \otimes \sigma^*) .$$

On the other hand, we know from Section 9 that $(i_V \otimes \sigma^*) \circ \rho^* = \rho^* \circ \rho^\circ(\sigma)$, whence $\exp(i_V \otimes \sigma^*) \circ \rho^* = \rho^* \circ \exp(\rho^\circ(\sigma))$. Hence our above expression for $\rho(\exp(\sigma))$ gives

$$\rho(\exp(\sigma)) = (i_V \otimes c) \circ \rho^* \circ \exp(\rho^\circ(\sigma)) = \exp(\rho^\circ(\sigma)) .$$

Hence $\rho(G) = \exp(\rho^\circ(\mathscr{L}(G)))$. The fact that, for the original ρ, the image $\rho(G)$ is an algebraic unipotent subgroup of H follows therefore from Theorem 10.1, so that Theorem 10.2 is now proved.

THEOREM 10.3. *Let (G, A) be the structure of a unipotent affine algebraic group over a field F of characteristic 0. Let H be a normal algebraic subgroup of G. Then $(G/H, A^H)$ is the structure of an affine algebraic* *group, and there is a polynomial map $G/H \to G$ whose composite with the canonical map $G \to G/H$ is the identity map on G/H.*

Proof: We know from Section 6 (remark just preceding Corollary 6.5) that A^H is finitely generated. Let ρ be the restriction morphism $G \to G(A^H)$. By Theorem 6.2, the kernel of ρ is precisely H. By Theorem 10.2, $\rho(G)$ is a unipotent algebraic subgroup of the affine algebraic group $\mathscr{G}(A^H)$. Let B denote the algebra of polynomial functions on $\rho(G)$. Clearly, $B \circ \rho \subset A^H$. We can evidently find a linear map τ: $\rho^\circ(\mathscr{L}(G)) \to \mathscr{L}(G)$ such that $\rho^\circ \circ \tau$ is the identity map on $\rho^\circ(\mathscr{L}(G))$. Let α denote the map $\exp \circ \tau \circ \log : \rho(G) \to G$. Clearly (in view of Theorem 10.1), α is a polynomial map, i. e., $A \circ \alpha \subset B$. Moreover, $\rho \circ \alpha$ is the identity map on $\rho(G)$. In fact, using Theorem 10.2, we obtain

$$\rho \circ \alpha = \rho \circ \exp \circ \tau \circ \log = \exp \circ \rho^\circ \circ \tau \circ \log = \exp \circ \log = i_{\rho(G)} .$$

Hence the map $f \to f \circ \rho$ of B into A^H is injective. Now let g be any element of A^H, and consider the element $g \circ \alpha$ of B. We have

$$(g \circ \alpha \circ \rho)(\exp(\sigma)) = (g \circ \alpha)(\exp(\rho^\circ(\sigma))) = g(\exp((\tau \circ \rho^\circ)(\sigma))) .$$

But $\sigma - (\tau \circ \rho^\circ)(\sigma)$ belongs to the kernel of ρ°, which is $\mathscr{L}(H)$, by Theorem 8.7. Hence it follows from the Campbell-Hausdorff formula that $\exp((\tau \circ \rho^\circ)(\sigma))$ belongs to $\exp(\sigma)H$. Since g lies in A^H, it follows that $g(\exp((\tau \circ \rho^\circ)(\sigma))) = g(\exp(\sigma))$. Since $G = \exp(\mathscr{L}(G))$, it follows now from the above that $g \circ \alpha \circ \rho = g$. Thus we have shown that the map $f \to f \circ \rho$ is an F-algebra isomorphism $B \to A^H$, its inverse being the map $g \to g \circ \alpha$. Moreover, if δ is any F-algebra homomorphism $A^H \to F$, then the map $f \to \delta(f \circ \rho)$ is an F-algebra homomorphism $B \to F$, so that there is an element x in G such that $\rho(x)(f) = \delta(f \circ \rho)$, i. e., such that δ is the restriction of x to A^H. Thus $(G/H, A^H)$ is indeed the structure of an affine algebraic group, isomorphic with $\rho(G)$. The required polynomial map $G/H \to G$ is the composite with $\alpha : \rho(G) \to G$ of the isomorphism $G/H \to \rho(G)$. This completes the proof of Theorem 10.3.

LEMMA 10.4. *Let G be a group of linear automorphisms of a finite-dimensional vector space. Let T be a normal unipotent subgroup of G, and let S be an arbitrary unipotent subgroup of G. Then ST is a unipotent subgroup of G.*

Proof: Let V denote the finite-dimensional vector space on which G acts. We view V as a $Z[G]$-module, where $Z[G]$ is the ordinary group algebra of G over the ring Z of integers. If s is an element of S, and t

is an element of T, we have, in $Z[G]$, $1 - st = s(1 - t) + (1 - s)$, and $(1 - t)s = s(1 - s^{-1}ts)$. Hence, a product $(1 - s_1t_1) \cdots (1 - s_nt_n)$ of such elements can be written as a sum of products of the form su, where s belongs to S, and u is a product whose factors are either of the form $1 - x$ with x in S, or of the form $1 - y$ with y in T, the total number of factors being n. We have

$$(1 - y)(1 - x) = (1 - y) - x(1 - x^{-1}yx) .$$

Hence each u can be written as an integral linear combination of products of the form

$$x(1 - x_1) \cdots (1 - x_p)(1 - y_1) \cdots (1 - y_q) ,$$

where x and the x_i's belong to S, and the y_j's belong to T. Moreover, q is the number of factors $1 - y$, with y in T, that occurred in the original expression for u. Therefore, if d is the dimension of V, the endomorphism that corresponds to u is 0 whenever $q \geqq d$. On the other hand, if $q < d$ and $n \geqq d^2$, then u must contain at least d successive factors of the form $1 - x$ with x in S, so that the corresponding endomorphism must again be 0. Thus the endomorphism corresponding to $(1 - s_1t_1) \cdots (1 - s_nt_n)$ is 0 whenever $n \geqq d^2$, so that Lemma 10.4 is established.

THEOREM 10.5. *Let G be an affine algebraic group over an arbitrary field. Let G_u denote the subgroup of G this is generated by the family of all normal unipotent subgroups of G. Then G_u is a normal unipotent algebraic subgroup of G.*

Proof: Let V be a finite-dimensional left stable subspace of $\mathscr{A}(G)$, and denote the dimension of V by d. We wish to show that the representation of G_u by left translations on V is unipotent, i. e., that a product $(1 - x_1^*) \cdots (1 - x_d^*)$ annihilates V whenever the x_i's belong to G_u. Since these x_i's are then already contained in the subgroup of G that is generated by a *finite* family of unipotent normal subgroups of G, we obtain this result by repeated applications of Lemma 10.2. Thus we conclude that G_u is locally unipotent on $\mathscr{A}(G)$, so that G_u is a unipotent subgroup of G. Clearly, G_u is normal in G. Finally, since the algebraic hull of a unipotent normal subgroup is still unipotent and normal, we have that G_u is an algebraic subgroup of G, so that Theorem 10.5 is proved.

The group G_u of Theorem 10.5 is called the *unipotent radical* of G. It plays an essential role in the structure theory of affine algebraic groups, as we shall see later on.

THEOREM 10.6. *Let G be a group of linear automorphisms of a finite-dimensional vector space V over an arbitrary field F. If every element of G is unipotent then the representation of G on V is unipotent.*

Proof: Clearly, no generality is lost in assuming that F is algebraically closed. Next, observe that if we proceed by induction on the dimension of V, we reduce the theorem to the case where V is simple as a G-module. In that case, let E be the algebra of linear endomophisms of V that is generated by the endomorphisms $x - i_V$, with x in G. Clearly, V is still simple as an E-module. Hence, if $E \neq (0)$, it follows from the classical Wedderburn structure theorem concerning associative algebras that E is the algebra of all linear endomorphisms of V. On the other hand, it is clear that every element of E is actually a linear combination of elements of the form $x - i_V$, with x in G. Since every such element is nilpotent by assumption, it follows that every element of E has trace 0. Hence E cannot contain every linear endomorphism of V. Thus we must have $E = (0)$, which proves Theorem 10.6.

COROLLARY 10.7. *Let G be an affine algebraic group over an arbitrary field F. If every element of G is unipotent then G is unipotent.*

Proof: This is an immediate consequence of Theorem 10.6.

EXERCISES

1. Let G be an affine algebraic group over a field F of characteristic $p \neq 0$. Show that every unipotent element of G (i. e., an element whose left translation action on $\mathcal{A}(G)$ is locally unipotent) is of finite order, its order being a power of p
2. Prove that every unipotent affine algebraic group is *nilpotent*, in the sense of ordinary group theory.
3. Let F be a field of characteristic 0, and let G be a commutative unipotent affine algebraic group over F. Show that G is isomorphic with a *vector group* over F, i. e., that $\mathcal{A}(G)$ is an ordinary polynomial algebra $F[t_1, \ldots, t_n]$, with comultiplication γ such that $\gamma(t_i) = 1 \otimes t_i + t_i \otimes 1$, and antipode η such that $\eta(t_i) = -t_i$, for each i.

11. ABELIAN GROUPS

Let (G, A) be the structure of an affine algebraic group over a perfect field F, and let e be any proper linear endomorphism of A. Then e stabilizes every left stable F-subspace of A. Using the fact that A is the union of the family of its finite-dimensional left stable F-subspaces, we see immediately that Theorem 1.15 extends so as to yield a Jordan decomposition of e, in the algebra of all proper linear endomorphisms of A. More precisely, there is one and only one pair $(e^{(s)}, e^{(n)})$ of proper linear endomorphisms of A such that $e = e^{(s)} + e^{(n)}$, $e^{(s)}e^{(n)} = e^{(n)}e^{(s)}$, $e^{(s)}$ is semisimple, and $e^{(n)}$ is locally nilpotent. Exactly as in the finite-dimensional case, this yields also a multiplicative decomposition $e = e^{(s)}e^{(u)}$, where $e^{(u)} = i_A + (e^{(s)})^{-1}e^{(n)}$, in the case where e is a proper linear automorphism of A. Here, $e^{(u)}$ is locally unipotent, but we shall call it simply the unipotent component of e. Similarly, we call $e^{(n)}$ the nilpotent component of e. The following result shows that these components are significant for the structure theory of affine algebraic groups. It will be convenient to extend the notation used above as follows. If σ is an element of the dual A° of A, and if $\sigma^* = (i_A \otimes \sigma) \circ \gamma$ is the corresponding proper linear endomorphism of A, then we write $\sigma^{(s)}$ for $c \circ \sigma^{*(s)}$, $\sigma^{(n)}$ for $c \circ \sigma^{*(n)}$, and $\sigma^{(u)}$ for $c \circ \sigma^{*(u)}$, where c is the counit of A. We refer to these as the semisimple, nilpotent, and unipotent components of σ, respectively.

THEOREM 11.1. *Let (G, A) be the structure of an affine algebraic group over a perfect field F. If x is an element of G, then the semisimple component $x^{(s)}$, and the unipotent component $x^{(u)}$, are elements of G. If σ is an element of $\mathscr{L}(G)$, then the semisimple component $\sigma^{(s)}$, and the nilpotent component $\sigma^{(n)}$, are elements of $\mathscr{L}(G)$. Moreover, if $\rho: G \to H$ is a morphism of affine algebraic groups, then $\rho(x^{(s)}) = \rho(x)^{(s)}$, $\rho(x^{(u)}) = \rho(x)^{(u)}$, $\rho^\circ(\sigma^{(s)}) = \rho^\circ(\sigma)^{(s)}$, and $\rho^\circ(\sigma^{(n)}) = \rho^\circ(\sigma)^{(n)}$.*

Proof: Let V be a finite-dimensional left stable F-subspace of A, and note that the F-linear combinations of products of pairs of elements of V constitute another finite-dimensional left stable F-subspace V^2 of A. Let τ be any linear automorphism of $V + V^2$ that stabilizes both V and V^2. We define a corresponding linear automorphism τ' of the space $\mathrm{Hom}_F(V \otimes V, V^2)$ by setting

$$\tau'(h) = \tau \circ h \circ (\tau_V^{-1} \otimes \tau_V^{-1})$$

for every element h of $\mathrm{Hom}_F(V \otimes V, V^2)$, where τ_V denotes the restriction of τ to V. It is easy to see that τ' is unipotent whenever τ is unipotent. We claim that τ' is semisimple whenever τ is semisimple. In order to see this, let F' be an algebraic closure of F. If τ is semisimple we have from Proposition 1.14 that the canonical extension of τ to an F'-linear automorphism of $(V + V^2) \otimes F'$ is still semisimple. We decompose the τ-stable subspaces $V \otimes F'$ and $V^2 \otimes F'$ into direct sums of 1-dimensional τ-stable subspaces, and then decompose the F'-space

$$\mathrm{Hom}_{F'}\big((V \otimes F') \otimes (V \otimes F'), V^2 \otimes F'\big) = \mathrm{Hom}_F(V \otimes V, V^2) \otimes F'$$

accordingly. This shows that the canonical F'-linear extension of τ' is semisimple. By Proposition 1.14, this implies that τ' is semisimple.

Now let x be an element of G, and let τ be the restriction of x^* to $V + V^2$. Since the components $\tau^{(s)}$ and $\tau^{(u)}$ are linear combinations of powers of τ, they stabilize V and V^2. By the above, $(\tau^{(u)})'$ is unipotent, and $(\tau^{(s)})'$ is semisimple. Clearly, these linear automorphisms commute with each other, and their product is τ'. Hence we must have $(\tau^{(u)})' = (\tau')^{(u)}$, and $(\tau^{(s)})' = (\tau')^{(s)}$.

Now the multiplication map $\mu : V \otimes V \to V^2$ is an element of $\mathrm{Hom}_F(V \otimes V, V^2)$, and we have $\tau'(\mu) = \mu$, owing to the fact that x^* is an F-algebra automorphism of A. The multiplicative Jordan components of the restriction of τ' to $F\mu$ are the restrictions to $F\mu$ of the multiplicative Jordan components of τ'. Hence $(\tau')^{(u)}$ and $(\tau')^{(s)}$ leave μ fixed, i.e., $(\tau^{(u)})'$ and $(\tau^{(s)})'$ leave μ fixed. Letting V range over the family of all finite-dimensional left stable F-subspaces of A, we conclude from this that $(x^*)^{(u)}$ and $(x^*)^{(s)}$ are F-algebra automorphisms of A. Since they are also proper, this means that $x^{(u)}$ and $x^{(s)}$ are elements of G.

The proof of the corresponding result for Lie algebra elements is almost the same. The only changes are the following. Instead of defining τ' as above, we define it now by

$$\tau'(h) = \tau \circ h - h \circ (i_V \otimes \tau_V + \tau_V \otimes i_V),$$

and we note that τ is an F-algebra derivation of A if and only if (for all V's as above) $\tau'(\mu) = 0$.

Finally, consider a morphism of affine algebraic groups $\rho: G \to H$. Let V be a finite-dimensional left stable F-subspace of $\mathscr{A}(H)$. Then ρ defines a polynomial representation of G on V. From the beginning of Section 9, we know that, as a G-module, V is isomorphic with a G-submodule of a direct sum of copies of a G-submodule of A. It is clear from this that, for every element x of G, $\rho(x^{(s)})^*$ is semisimple on V, and $\rho(x^{(u)})^*$ is unipotent on V. Since this is true for all V's, we have that $\rho(x^{(s)})^*$ is semisimple, and that $\rho(x^{(u)})^*$ is locally unipotent. Since they commute with each other, these are therefore the multiplicative Jordan components of $\rho(x)^*$. Hence we must have $\rho(x^{(s)}) = \rho(x)^{(s)}$, and $\rho(x^{(u)}) = \rho(x)^{(u)}$. The proof of the corresponding result for the differential ρ° of ρ is exactly the same, so that Theorem 11.1 is now proved.

In this section, we are primarily interested in direct product decompositions of abelian affine algebraic groups. Later on, we shall be concerned with semidirect product decompositions of non-commutative groups. In order to be prepared for this, we shall discuss the generalities concerning semidirect products here.

Let (G, A) be the structure of an affine algebraic group over an arbitrary field F, and let N be a normal algebraic subgroup of G. In the case where F is algebraically closed, we know from Corollary 6.5 that $(G/N, A^N)$ is the structure of an affine algebraic group. This is not always true when F is not algebraically closed. In fact, there are examples (where F is the field of real numbers) in which there is no structure of affine algebraic group on G/N such that the canonical map $G \to G/N$ is a morphism of affine algebraic groups. We shall say that N is a *properly normal* algebraic subgroup of G if $(G/N, A^N)$ is the structure of an affine algebraic group. By Theorem 6.2 and the remark immediately preceding Corollary 6.5, N is *properly* normal if and only if every F-algebra homomorphism $A^N \to F$ is the restriction of an element of G.

Suppose that we are given a properly normal algebraic subgroup N of G, and an algebraic subgroup P of G, such that the following conditions are satisfied: (1) $G = NP$; (2) $N \cap P$ is trivial; (3) the restriction to P of the canonical map $G \to G/N$ is an isomorphism of affine algebraic groups (it is clearly a bijective morphism; the extra assumption is that its inverse be also a morphism of affine algebraic groups). Under these circumstances, we say that G is the *semidirect product* of N and P.

Such a semidirect product decomposition $G = N \cdot P$ leads to a tensor product decomposition of A, as follows. The restriction images A_P and A_N are the algebras of polynomial functions on P and N, respectively. By assumption (3), the injection $A^N \to A$, followed by the restriction map $A \to A_P$, is an isomorphism of F-algebras. Let $\rho : A_P \to A^N$ denote its inverse. The canonical map $G \to G/N$, followed by the inverse $G/N \to P$ of its restriction to P, is a morphism $\pi : G \to P$ of affine algebraic groups. From this, we define a polynomial map $\pi' : G \to N$ by putting $\pi'(x) = x\pi(x)^{-1}$ for every x in G (if $x = np$, with n in N and p in P, then $\pi'(x) = n$). Now, if f is any element of A_N, then $f \circ \pi'$ belongs to the left P-fixed part A^P of A, and this defines an F-algebra homomorphism, $\sigma : A_N \to A^P$. One sees immediately that σ is an isomorphism, its inverse being the restriction map $A^P \to A_N$.

Now we can show that the multiplication map $A^P \otimes A^N \to A$ is an isomorphism. Let f be any element of A, let n be an element of N, and let p be an element of P. We write $\gamma(f) = \sum_i u_i \otimes v_i$, with each u_i and v_i in A, so that $f(np) = \sum_i u_i(n)v_i(p)$. We see directly from the definitions of ρ and σ that $v_i(p) = \rho\big((v_i)_P\big)(np)$, and $u_i(n) = \sigma\big((u_i)_N\big)(np)$. Hence $f = \sum_i \sigma\big((u_i)_N\big)\rho\big((v_i)_P\big)$, which shows that the above multiplication map is surjective. From the fact that $N \times P$ separates the elements of $A_N \otimes A_P$, and therefore also separates the elements of the isomorphic F-algebra $A^P \otimes A^N$, it follows that this map is also injective.

A subgroup P of G is said to be *A-semisimple*, or *reductive*, if the representation of P by left translations on A is semisimple (this implies, through the intervention of the antipode, that the representation of P by right translations on A is also semisimple). An equivalent condition is that the restriction to P of every polynomial representation of G be semisimple, as is seen immediately from our discussion at the beginning of Section 9. By a well-known elementary result from the general representation theory of groups, the same is then true for every normal subgroup of P. If $[P]$ denotes the algebraic hull of P in G, then the $[P]$-stable subspaces of A are easily seen to coincide with the P-stable subspaces. Hence *P is a reductive subgroup of G if an only if its algebraic hull is a reductive subgroup of G.*

PROPOSITION 11.2 *Let (G, A) be the structure of an affine algebraic group over an arbitrary field F. Suppose that P is a reductive subgroup of G such that $G_uP = G$. Then P is an algebraic subgroup of G, and G is the semidirect product of G_u and P.*

Proof: If $[P]$ is the algebraic hull of P in G, then it is clear from the remarks above that $G_u \cap [P]$ is reductive, as well as unipotent. Hence

$G_u \cap [P]$ is trivial. Since $G_u P = G$, we see from this that $P = [P]$, and that there is one and only one map $\pi : G \to P$ such that $x\pi(x)^{-1} \in G_u$ for every x in G. Clearly, π is a group homomorphism, and we shall prove that π is actually a morphism of affine algebraic groups, i.e., that $A_P \circ \pi \subset A$ or, equivalently, that $A \circ \pi \subset A$. Let V be a finite-dimensional left stable F-subspace of A that generates A as an F-algebra. Let $(0) = V_n \subset \cdots \subset V_0 = V$ be composition series for V as a G-module, and let V' denote the direct sum of the simple factor G-modules V_i/V_{i+1}. Since V is semisimple as a P-module, there is a P-module isomorphism $\delta : V' \to V$. For every element x of G, we denote by x_V and $x_{V'}$ the corresponding F-linear automorphisms of V and V', respectively. Since G_u acts trivially on each V_i/V_{i+1}, we have $x_{V'} = \pi(x)_{V'}$. Since δ is a P-module isomorphism, we have $\pi(x)_V \circ \delta = \delta \circ \pi(x)_{V'}$. Hence $\pi(x)_V = \delta \circ x_{V'} \circ \delta^{-1}$. Now let v be any element of V. Then $(v \circ \pi)(x) = v(\pi(x)) = c(\pi(x)_V(v))$, where c is the co-unit of A. By the above, this is equal to $c((\delta \circ x_{V'} \circ \delta^{-1})(v))$. In particular, this shows that $v \circ \pi$ is a representative function associated with the polynomial representation G-module V', so that $v \circ \pi$ belongs to A. Since V generates A as an F-algebra, it follows that $A \circ \pi \subset A$, so that π is indeed a morphism of affine algebraic groups.

Now let π' denote the map of G into G_u that is defined by $\pi'(x) = x\pi(x)^{-1}$. Since π is a polynomial map, so is π', i.e., $A \circ \pi' \subset A$. Moreover, it is clear from the definition of π' that each $a \circ \pi'$ is left P-fixed. Thus we actually have $A \circ \pi' \subset A^P$. Similarly, we see that $A \circ \pi \subset A^{G_u}$ (noting that G_u is normal in G). Now it follows, exactly as in our above discussion of semidirect products, that the multiplication map $A^P \otimes A^{G_u} \to A$ is surjective. Since $G_u P = G$, it is clear that G_u separates the elements of A^P, and that P separates the elements of A^{G_u}. Hence $G_u \times P$ separates the elements of $A^P \otimes A^{G_u}$, whence we see that the multiplication map $A^P \otimes A^{G_u} \to A$ is also injective, so that it is an F-algebra isomorphism. It is clear from this that every F-algebra homomorphism $A^{G_u} \to F$ is the restriction of an element of G. Thus G_u is a properly normal algebraic subgroup of G. Finally, our above morphism π has kernel G_u, and therefore induces a morphism of affine algebraic groups $G/G_u \to P$ in the natural way. This is clearly an isomorphism, its inverse being the restriction to P of the canonical morphism $G \to G/G_u$. Proposition 11.2 is therefore established.

The simplest application of Proposition 11.2 concerns the structure of abelian groups over a perfect field. The result is the following.

THEOREM 11.3. *Let G be an abelian affine algebraic group over a perfect field. Then there is one and only one $\mathscr{A}(G)$-semisimple algebraic subgroup G_s of G such that G is the direct product $G_u \times G_s$, where G_u is the unipotent radical of G. The group G_s consists precisely of all $\mathscr{A}(G)$-semisimple elements of G.*

Proof: Let F' be an algebraic closure of the base field F, and consider the extended group $G^{F'}$, regarding G as a subgroup of $G^{F'}$. By Proposition 1.14, every $\mathscr{A}(G)$-semisimple element x of G is thus identified with an $\mathscr{A}(G^{F'})$-semisimple element of $G^{F'}$. Since F' is algebraically closed, $\mathscr{A}(G^{F'})$ is the direct sum of a family of one-dimensional x^*-stable subspaces. Since $G^{F'}$ is abelian, the partial sums corresponding to the characteristic values of x^* are left stable. Hence it is clear that the $\mathscr{A}(G)$-semisimple elements of G constitute an $\mathscr{A}(G^{F'})$-semisimple subgroup of $G^{F'}$. By Proposition 1.14, this implies that they constitute also an $\mathscr{A}(G)$-semisimple subgroup of G. Let G_s be this subgroup of G. Clearly, the algebraic hull of G_s is still an $\mathscr{A}(G)$-semisimple subgroup of G. Since it is abelian, this implies that all its elements are $\mathscr{A}(G)$-semisimple, so that it must coincide with G_s. Thus G_s is an algebraic subgroup of G. By Theorem 11.1, every element x of G is a product $x^{(u)}x^{(s)}$, with $x^{(u)}$ being $\mathscr{A}(G)$-unipotent, and $x^{(s)}$ being $\mathscr{A}(G)$-semisimple. Clearly, $x^{(u)}$ belongs to G_u, and $x^{(s)}$ belongs to G_s, so that x belongs to G_uG_s. Now we can apply Proposition 11.2 and conclude that G is the direct product $G_u \times G_s$. Since every element of an $\mathscr{A}(G)$-semisimple subgroup of G is $\mathscr{A}(G)$-semisimple, it is clear that G_s must contain every $\mathscr{A}(G)$-semisimple subgroup of G. Hence G_s is the only $\mathscr{A}(G)$-semisimple subgroup of G whose product with G_u is G. This completes the proof of Theorem 11.3.

Recall that the multiplicative group of the non-zero elements of a field F is an affine algebraic group F^*, with $\mathscr{A}(F^*)$ the F-algebra generated by u and its reciprocal u^{-1}, where $u(x) = x$ for every non-zero element x of F. The comultiplication γ of $\mathscr{A}(F^*)$ is determined by $\gamma(u) = u \otimes u$. If F is an infinite field, then u is not algebraic over F, and we have $\mathscr{A}(F^*) = F[u, u^{-1}]$, with u playing the role of an independent variable in the F-algebra structure of $\mathscr{A}(F^*)$. A direct product of a finite family of copies of this affine algebraic group F^* is called a *toroid*. Thus, if T is a toroid over an infinite field F, then

$$\mathscr{A}(T) = F[u_1, \ldots, u_n, u_1^{-1}, \ldots, u_n^{-1}],$$

THEOREM 11.4. *Every connected abelian reductive affine algebraic group over an algebraically closed field is a toroid.*

Proof: Let G be such a group, let $A = \mathscr{A}(G)$, and let F denote the base field. Choose a finite-dimensional left stable F-subspace V of A that generates A as an F-algebra. Since G is reductive, i.e., A-semisimple, V is the sum of the family of its G-simple subspaces. Since G is abelian. it acts on each simple G-submodule of V by scalar multiplications. Thus each simple G-submodule W of V determines a group homomorphism $\tau_W : G \rightarrow F^*$ such that every element x of G acts on W by the scalar multiplication with $\tau_W(x)$. Since V generates A as an F-algebra, it is clear that the group U generated by these homomorphisms τ_W consists precisely of all the group homomorphisms $G \rightarrow F^*$ that are associated with simple G-submodules of A. We know that U is contained in A, and that A consists precisely of all the representative functions on G that are associated with G-submodules of A. Hence it is clear that U generates A as an F-algebra. Moreover, since the above V is the sum of a *finite* family of W's as above, we know that U is a *finitely generated* abelian group. If u is any element of U then $u(G)$ is a connected algebraic subgroup of F^*, by Corollary 5.2. Since F^* is one-dimensional, it follows that u is either trivial or surjective (Theorem 7.4 and Proposition 8.5). Clearly, this implies that the group U is torsion-free. Hence U has a finite free basis $(u_1, \ldots, u_n)$. By a well-known elementary result, the elements of U are F-linearly independent. Hence the set $(u_1, \ldots, u_n)$ is algebraically independent over F. Since

$$\mathscr{A}(G) = F[U] = F[u_1, \ldots, u_n, u_1^{-1}, \ldots, u_n^{-1}],$$

it is clear from our above discussion of toroids that G is a toroid, so that Theorem 11.4 is proved.

THEOREM 11.5. *Let G be a connected abelian affine algebraic group over an algebraically closed field, and let K be an algebraic subgroup of G. Suppose that τ is a morphism of affine algebraic groups $K \rightarrow T$, where T is a toroid. Then τ extends to a morphism of affine algebraic groups $G \rightarrow T$.*

Proof: It is evidently sufficient to prove the theorem in the case where $T = F^*$. By Theorem 11.3, we have the direct product decompositions $G = G_u \times G_s$, and $K = K_u \times K_s$. Evidently, K_s is an algebraic subgroup of G_s, and K_u is an algebraic subgroup of G_u. Since G is connected, so is G_s. By Theorem 11.4, G_s is therefore a toroid. Since $\tau(K_u)$ is a uni-

potent subgroup of the reductive group F^*, it is clear that K_u is contained
where the u_i's are independent variables over F, and $\gamma(u_i) = u_i \otimes u_i$ for
each i. If η is the antipode of $\mathscr{A}(F^*)$ we have $\eta(u_i) = u_i^{-1}$ for each i.
in the kernel of τ. Therefore, it suffices to prove that the restriction of τ to K_s extends to a morphism $G_s \to F^*$. Thus we may now assume that G is a toroid.

Let us regard τ as an element of $\mathscr{A}(K)$, and let us choose an element f of $\mathscr{A}(G)$ such that the restriction f_K is τ. Since G is a toroid, f is an F-linear combination $\sum_{i=1}^{m} r_i h_i$, where the r_i's are elements of F, and the h_i's are morphisms of affine algebraic groups $G \to F^*$. We choose f so that m is as small as possible. Now the F-linear combination $\sum_{i=1}^{m} r_i(h_i)_K$ of the group homomorphisms $(h_i)_K : K \to F^*$ is the group homomorphism τ. Since m is minimal, no two $(h_i)_K$'s, are equal. If τ were not equal to one of the $(h_i)_K$'s, the set $((h_1)_K, \ldots, (h_m)_K, \tau)$ would therefore be linearly independent. Hence we must have $m = 1$ and $\tau = (h_1)_K$. Thus h_1 is the required extension of τ, and Theorem 11.5 is proved.

COROLLARY 11.6. *Let G be a connected abelian affine algebraic group over an algebraically closed field. Suppose that a toroid T is an algebraic subgroup of G. Then T is a direct factor of G.*

Proof: By Theorem 11.5, we have a morphism of affine algebraic groups $\rho : G \to T$ that extends the identity map $T \to T$. Let P be the kernel of ρ, and consider the product map $T \times P \to G$. This is evidently a bijective morphism of affine algebraic groups. Its inverse sends each element x of G onto $(\rho(x), \rho(x)^{-1}x)$, and is evidently a morphism of affine algebraic groups. Hence G is the direct product $T \times P$.

EXERCISES

1. Let (G, A) be the structure of an affine algebraic group over the field F. The elements of A that are group homomorphisms $G \to F^*$ evidently constitute a multiplicative group $X(G)$, called the *character group* of G. Show that if F is infinite and G is a toroid over F then $X(G)$ is a finitely generated free abelian group.
2. Let F be an algebraically closed field, and let S be a finitely generated abelian group. Make the ordinary group algebra $F[S]$ into a Hopf algebra, defining the comultiplication γ and the antipode η so that $\gamma(s) = s \otimes s$ and $\eta(s) = s^{-1}$ for every element s of S. In the case where F is of non-zero characteristic p, assume that S has no element of order p. Prove that then $F[S]$ has no nilpotent elements other

than 0, hence that $(\mathscr{G}(F[S]), F[S])$ is the structure of an abelian reductive affine algebraic group, and that its character group (as defined in Exercise 1) may be identified with S.

3. Let G be a reductive abelian affine algebraic group over an algebraically closed field F. Show that the character group $X(G)$ of G is a group S as in Exercise 2, and that the natural map $G \to \mathscr{G}(F[X(G)])$ is an isomorphism of affine algebraic groups.

12. SEMISIMPLE REPRESENTATIONS

Let us consider the following general setting: G is a group, F a field, B a fully stable F-algebra of F-valued representative functions on G. By a *B-representation* of G, we mean a representation of G by linear automorphisms of a finite-dimensional F-space whose associated representative functions belong to B. As we have already remarked in Section 11, it is an elementary matter to show that *the restriction to a normal subgroup K of G of a finite-dimensional semisimple representation of G is always a semisimple representation of K.* The following is an important result in the converse direction. We use the notation just introduced, and we let B_K denote the restriction image of B in the F-algebra of representative functions on K.

PROPOSITION 12.1. *Suppose that every B-representation of G whose kernel contains K is semisimple, and that the tensor product of semisimple B_K-representation spaces of K is always semisimple. Then every B-representation space V of G that is semisimple with respect to K is also semisimple with respect to G.*

Proof: Let W be a G-stable subspace of V, and let n be the dimension of W over the base field F. For every F-space S, let S^n stand for the homogeneous component of degree n of the exterior F-algebra built over S. The G-module structure of V is canonically extended (via tensor products) to a G-module structure of V^n, with which V^n is evidently a B-representation space of G. Clearly, W^n is a one-dimensional G-stable subspace of V^n. Hence there is an element f in B, actually a group homomorphism of G into F^*, such that, for every element u of W^n, and every element x of G, the transform $x \cdot u$ is the scalar multiple $f(x)u$ of u. Our assumption on K implies (because V^n is a K-module homomorphic image of the n-th tensor power of V) that V^n is semisimple with respect to K. It is easily seen from this that there is a direct K-module decomposition $V^n = P + Q$, where P consists of all elements p such that $x \cdot p = f(x)p$ for every element x of K, and Q consists of all sums of elements of the form $x \cdot a - f(x)a$, with x in K and a in V^n. Using the fact that f is a homomorphism, and that K is normal in G, one verifies immediately that both P and Q are actually G-stable.

Now we define a new representation of G on P by setting $x(p) = f(x^{-1})x \cdot p$ for every x in G, and every p in P. Evidently, this is again a B-representation of G, and its kernel contains K. By assumption, this is therefore a semisimple representation of G. Hence the G-stable subspace W^n of P has a G-module complement, R say, in P, with respect to our new representation. However, it is clear from the definition that the G-stable subspaces of P for our new representation are the same as the G-stable subspaces for the original action of G on P. Thus R is G-stable also for the original G-module structure of V^n, and we have a direct G-module decomposition $V^n = W^n + R + Q$. In particular, W^n has a G-module complement $R + Q = S$, say, in V^n.

Let W_1 denote the F-subspace of V consisting of all those elements v for which the exterior product vW^{n-1} is contained in S. Since S and W^{n-1} are G-stable, so is W_1. Now observe that S is of codimension 1 in V^n, so that S is the space of zeros of some linear functional μ on V^n. The space W^{n-1} is of dimension n, and if $(u_1, \ldots, u_n)$ is a basis of W^{n-1} then an element v of V belongs to W_1 if and only if $\mu(vu_i) = 0$ for each i. Hence the dimension of W_1 is at least $m - n$, where m is the dimension of V. Now let w_1 be any non-zero element of W, and choose elements $w_2, \ldots, w_n$ such that $(w_1, \ldots, w_n)$ is a basis of W. Then $w_2 \cdots w_n$ belongs to W^{n-1}, and $w_1 \cdots w_n$ is a non-zero element of W^n. Hence w_1 doesn ot belong to W_1, and we conclude that $W \cap W_1 = (0)$, which implies that the dimension of W_1 is at most $m - n$. From the above, we have therefore that W_1 is of dimension $m - n$ exactly. Hence V is the direct sum of the G-stable subspaces W and W_1. Thus we have shown that every G-submodule of V has a G-module complement in V, so that V is semisimple with respect to G. This establishes Proposition 12.1.

THEOREM 12.2. *Let U and V be finite-dimensional semisimple representation spaces, over a field of characteristic 0, for an arbitrary group G. Then the tensor product G-module $U \otimes V$ is semisimple.*

Proof: Let S denote the algebraic hull of the image of G in the affine algebraic group $G(U + V)$ of all linear automorphisms of the direct sum $U + V$. Clearly, S stabilizes U and V, and the S-stable subspaces of U and V coincide with the G-stable subspaces, Moreover, it is clear that the images of S and G in $G(U \otimes V)$ have the same algebraic hull, so that the S-stable subspaces of $U \otimes V$ coincide with the G-stable subspaces of $U \otimes V$. Hence it suffices to prove Theorem 12.2 for the S-modules U and V. This means that we may assume that G is an affine algebraic group, and that the representations of G on U and V are polynomial representations.

Assuming this, let G_1 denote the connected component of the neutral element in G. Since G/G_1 is finite, and since the base field is of characteristic 0, we know that every representation of G/G_1 over our base field is semisimple. Hence Theorem 12.2 will follow immediately from Proposition 12.1 as soon as we have proved it in the case where G is connected. Accordingly, let us now assume that G is connected. Then it is clear from Theorem 9.1 that, for every polynomial representation of G, the G-stable subspaces coincide with the $\mathscr{L}(G)$-stable subspaces. Hence it suffices to prove that, if L is any finite-dimensional Lie algebra over a field of characteristic 0, and if U and V are finite-dimensional semisimple L-modules, then the tensor product L-module $U \otimes V$ is semisimple. In doing this, we may evidently replace L with its image in the Lie algebra of all linear endomorphisms of $U + V$. Thus we may assume that L is given to us as a Lie algebra of linear endomorphisms of $U + V$ stabilizing U and V, and that $U + V$ is semisimple as an L-module.

Then it follows (from a standard theorem on representations of Lie algebras, due to N. Jacobson) that the commutator Lie subalgebra $[L, L]$ is semisimple, that L is the direct sum of its center, T say, and $[L, L]$, and that every element of T is a semisimple linear endomorphism of $U + V$. By Proposition 1.14, no generality is lost in assuming that the base field is algebraically closed, which we shall now suppose to be the case. Then U and V are direct sums of one-dimensional T-stable subspaces, to each of which there corresponds a linear functional μ on T such that every element t of T acts as the scalar multiplication by $\mu(t)$ on that one-dimensional subspace. Adding together all those one-dimensional T-stable subspaces of U or of V for which the linear functionals μ coincide, we obtain direct L-module decompositions $U = U_1 + \cdots + U_p$, an $V = V_1 + \cdots + V_q$, such that T acts on each U_i via a certain linear functional σ_i, and on each V_j via a certain linear functional τ_j. Now it suffices to show that each $U_i \otimes V_j$ is semisimple as an L-module.

Since $[L, L]$ is a semisimple Lie algebra, $U_i \otimes V_j$ is semisimple as an $[L, L]$-module. On the other hand, T acts on $U_i \otimes U_j$ by scalar multiplication, via the linear functional $\sigma_i + \tau_j$. Hence the L-stable subspaces of $U_i \otimes V_j$ coincide with the $[L, L]$-stable subspaces. Thus $U_i \otimes V_j$ is semisimple as an L-module, and our proof of Theorem 12.2 is now complete.

The next theorem is basic for the representation and structure theory of solvable groups. It is usually stated by saying that, over an algebraically closed field, a connected solvable linear algebraic group can be put in triangular (matrix) form.

THEOREM 12.3 (Lie-Kolchin). *Let G be a connected solvable affine algebraic group over an algebraically closed field F. Then every simple polynomial representation of G is 1-dimensional.*

Proof: Let V be a simple (non-zero) polynomial representation space for G, and let K denote the commutator subgroup of G. By Corollary 5.3, K is a connected algebraic subgroup of G. Since G is solvable, the dimension of K is strictly smaller than that of G, unless G is trivial. Making an induction on the dimension of G, we may therefore suppose that Theorem 12.3 holds for K. Since K is normal in G, we know that V is semisimple as a K-module. Hence our inductive hypothesis implies that V is the direct sum of 1-dimensional K-stable subspaces. As in our proof of Theorem 12.2, this leads to a direct K-module decomposition $V = V_1 + \cdots + V_p$ with corresponding morphisms of affine algebraic groups $\mu_i : K \to F^*$ such that every element x of K acts on each V_i as the scalar multiplication by $\mu_i(x)$, the μ_i's being mutually distinct. From the fact that K is normal in G, we see that every element of G permutes the V_i's among themselves. Hence the stabilizer of V_1 in G is of finite index in G. Since G is connected, it follows that V_1 is G-stable. Since V is simple, we have therefore $V = V_1$, which means that K acts by scalar multiplication on V. Since K is the commutator subgroup of G, the determinant of the linear automorphism effected by an element of K on V is equal to 1. Hence, if n is the dimension of V, we have $\mu_1(x)^n = 1$ for every element x of K. Since K is connected, it follows that $\mu_1(x) = 1$ for every x in K, i.e., that K acts trivially on V. Thus we may view V as a module for the abelian factor group G/K. By Schur's lemma, the simplicity of V therefore implies that V is 1-dimensional, so that Theorem 12.3 is proved.

EXERCISES

1. Let F be a field of non-zero characteristic p. Let V be a p-dimensional vector space over F, and let G be the group of all linear automorphisms of V. Let $T^p(V)$ denote the homogeneous component of degree p of the tensor algebra built over V. Prove that the p-fold tensor product representation of G on $T^p(V)$ is not semisimple, thus showing that the assumption that F be of characteristic 0 is needed for the validity of Theorem 12.2. (Let J denote the kernel of the canonical homomorphism of $T^p(V)$ into the exterior algebra built over V. Note that J is a G-submodule of $T^p(V)$, that $T^p(V)/J$ is 1-dimensional, and that every element x of G acts on it as the scalar multiplication by the determinant $d(x)$ of x. Now show that J cannot have a G-module complement in $T^p(V)$).
2. Let G be an affine algebraic group over a field F of characteristic 0. Suppose that G has a semisimple representation whose associated representative functions generate $\mathcal{A}(G)$ as an F-algebra. Prove that G is reductive.

13. ALGEBRAIC LIE ALGEBRAS

Let G be an affine algebraic group over a field F. A Lie subalgebra L of $\mathscr{L}(G)$ is called an *algebraic Lie subalgebra* if there is an algebraic subgroup H of G such that $L = \mathscr{L}(H)$. In the case where F is of characteristic 0, we see immediately from Theorem 8.6 that, for every Lie subalgebra L of $\mathscr{L}(G)$, there is a unique smallest algebraic Lie subalgebra containing L, namely $\mathscr{L}(G_L)$. We call this the *algebraic hull* of L in $\mathscr{L}(G)$, and we denote it by $[L]$. *Throughout the remainder of this section, we assume that our base field F is of characteristic* 0.

Let M be a finite-dimensional polynomial representation space for G, and let U and V be F-subspaces of M such that $U \subset V$. The elements x of G with the property that $x \cdot v - v$ belongs to U for every element v of V evidently constitute an algebraic subgroup H of G. It is clear from Theorem 9.1 that $\mathscr{L}(H)$ consists precisely of those elements of $\mathscr{L}(G)$ which map V into U. Hence, *if L is any Lie subalgebra of $\mathscr{L}(G)$ that maps V into U, then $[L]$ also maps V into U.* We shall use this in proving the next two results.

Proposition 13.1. *Let G be an affine algebraic group over a field of characteristic* 0, *and let L be any Lie subalgebra of $\mathscr{L}(G)$. Then*

$$\big[[L], [L]\big] = [L, L].$$

Proof: Consider the adjoint representation of G on $\mathscr{L}(G)$. In the above, let $M = \mathscr{L}(G)$, $U = [L, L]$, and $V = L$. We conclude immediately that $\big[[L], L\big] \subset [L, L]$. Now let $U = [L, L]$, and $V = [L]$. Then the above gives $\big[[L], [L]\big] \subset [L, L]$, so that $\big[[L], [L]\big] = [L, L]$. This proves Proposition 13.1.

Proposition 13.2. *Let G be an affine algebraic group over a field of characteristic* 0, *and let L be a semisimple Lie subalgebra of $\mathscr{L}(G)$. Then L is an algebraic Lie subalgebra.*

Proof: Since every finite-dimensional representation of L is semisimple, the algebra A of polynomial functions on G is semisimple as an L-module, with respect to the action of L by proper derivations of A. We express this fact by saying that L is A-semisimple. Now we may apply the remark preceding Proposition 13.1 to each finite-dimensional left stable subspace of A in order to conclude that $[L]$ is still A-semisimple. By Jacobson's theorem (cf. proof of Theorem 12.2), this implies that $[L]$ is the direct sum of its center, T, say, and $\big[[L], [L]\big]$, and that every element of T is A-semisimple. From Proposition 13.1 we know that $\big[[L], [L]\big] = [L, L]$. Hence it suffices to show that $T = (0)$.

There is a finite-dimensional left stable subspace V of A such that the representation of $[L]$ on V is faithful. Let σ be an element of T, and let r be a characteristic root of the restriction of σ^* to V. Adjoining r to our base field F, let $W = V \otimes F[r]$, and extend the $[L]$-module structure of V canonically to an $[L]$-module structure of W. Let W_r denote the characteristic subspace of W that corresponds to r. Then W_r is evidently an $[L]$-submodule of W. Let d denote the dimension of W_r over $F[r]$, and consider the homogeneous component W^d of degree d in the exterior $F[r]$-algebra built over W. The G-module structure of W that is obtained by $F[r]$-linear extension from the G-module structure of V defines a polynomial representation of G on W^d, whose differential is the $\mathscr{L}(G)$-module structure of W^d that is obtained canonically from the $\mathscr{L}(G)$-module structure of V, via W. Now let $(w_1, \ldots, w_d)$ be an $F[r]$-basis of W_r. Then the transform of the exterior product $w_1 \cdots w_d$ by σ is equal to $(dr)w_1 \cdots w_d$. On the other hand, the space spanned by $w_1 \cdots w_d$ is $[L]$-stable, because W_r is $[L]$-stable. Therefore, $w_1 \cdots w_d$ is annihilated by $\big[[L], [L]\big] = L$. The same argument we used several times above shows that $w_1 \cdots w_d$ is therefore annihilated also by $[L]$. Hence we find that $dr = 0$, whence $r = 0$. Now we have shown that 0 is the only characteristic root of the restriction of σ^* to V. Since σ^* is semisimple, and since the representation of $[L]$ on V is faithful, it follows that $\sigma = 0$. Thus $T = (0)$, and Proposition 13.2 is proved.

THEOREM 13.3. *Let G be a connected affine algebraic group over a field F of characteristic 0. Let $\big[[G, G]\big]$ denote the algebraic hull of the commutator subgroup $[G, G]$ of G. Then $\mathscr{L}\big(\big[[G, G]\big]\big) = [\mathscr{L}(G), \mathscr{L}(G)]$.*

Proof: First, let us suppose that F is algebraically closed. Let R denote the radical of $\mathscr{L}(G)$ (i. e., its unique maximum solvable ideal). By Levi's theorem, there is a semisimple Lie subalgebra S of $\mathscr{L}(G)$ such that $\mathscr{L}(G) = S + R$. Hence $[\mathscr{L}(G), \mathscr{L}(G)] = S + T$, where T is the ideal $[\mathscr{L}(G), R]$ of $\mathscr{L}(G)$. Since R is a solvable ideal of $\mathscr{L}(G)$, it follows from standard Lie algebra theory (the theorem of Lie and Engel) that the restriction to T of every finite-dimensional representation of $\mathscr{L}(G)$ is nilpotent. In particular, T is therefore A-nilpotent, where $A = \mathscr{A}(G)$. Hence we have from Theorem 10.1 that T is an algebraic Lie subalgebra of $\mathscr{L}(G)$; $T = \mathscr{L}(G_T)$. By Proposition 13.2, S is an algebraic Lie subalgebra of $\mathscr{L}(G)$; $S = \mathscr{L}(G_S)$. From the fact that T is an ideal of $\mathscr{L}(G)$, it follows that G_T is normal in G, by Theorem 9.5. Hence $G_S G_T$ is a subgroup of G. Since F is algebraically closed, we have from Corollary 5.4 that $G_S G_T$ is a connected algebraic subgroup of G. The intersection $G_S \cap G_T$ is unipotent, because it is a subgroup of the unipotent group G_T. On the other hand, it is a normal subgroup of the A-semisimple group G_S, and is therefore A-semisimple. Hence $G_S \cap G_T$ must be trivial. Consider the canonical morphism of affine algebraic groups $G_S G_T \to (G_S G_T)/G_T$. Since $G_S \cap G_T$ is trivial, the restriction to G_S of this morphism is injective, and is therefore a bijective morphism $G_S \to (G_S G_T)/G_T$. By the note at the end of Section 6, this is an isomorphism of affine algebraic groups. Hence $\mathscr{L}((G_S G_T)/G_T)$ is isomorphic with $\mathscr{L}(G_S) = S$. On the other hand, we know from Theorem 8.7 that the kernel of the differential of the morphism $G_S G_T \to (G_S G_T)/G_T$ is $\mathscr{L}(G_T) = T$. By Theorem 7.5, this differential is surjective. Hence we conclude that the dimension of $\mathscr{L}(G_S G_T)$ is equal to the sum of the dimensions of S and T. Clearly, $S + T \subset \mathscr{L}(G_S G_T)$. Since $S \cap T = (0)$, our dimension result shows that $S + T = \mathscr{L}(G_S G_T)$, i.e., $[\mathscr{L}(G), \mathscr{L}(G)] = \mathscr{L}(G_S G_T)$. In particular, $[\mathscr{L}(G), \mathscr{L}(G)]$ is an algebraic Lie subalgebra of $\mathscr{L}(G)$, so that it coincides with $\mathscr{L}(G_{[\mathscr{L}(G), \mathscr{L}(G)]})$.

Appealing to Theorems 7.5 and 8.7, as above, we see that the Lie algebra of $G/G_{[\mathscr{L}(G), \mathscr{L}(G)]}$ is isomorphic with $\mathscr{L}(G)/[\mathscr{L}(G), \mathscr{L}(G)]$, which is abelian. By Theorem 9.4, $G/G_{[\mathscr{L}(G), \mathscr{L}(G)]}$ is therefore abelian, so that $[G, G] \subset G_{[\mathscr{L}(G), \mathscr{L}(G)]}$. By Theorem 8.6, we have therefore $\mathscr{L}([G, G]) \subset [\mathscr{L}(G), \mathscr{L}(G)]$ (by Corollary 5.3, $[G, G]$ is a connected algebraic subgroup of G).

In the general case, let F' denote an algebraic closure of F. Then we know from what we have proved above that $\mathscr{L}([G^{F'}, G^{F'}])$ is contained in $[\mathscr{L}(G^{F'}), \mathscr{L}(G^{F'})]$. By Proposition 7.1, we have $\mathscr{L}(G^{F'}) = \mathscr{L}(G) \otimes F'$, so that $\mathscr{L}([G^{F'}, G^{F'}]) \subset [\mathscr{L}(G), \mathscr{L}(G)] \otimes F'$. Clearly, $[[G, G]]^{F'} \subset [G^{F'}, G^{F'}]$. Using Theorem 8.6 and Proposition 7.1, we deduce from the last two relations that $\mathscr{L}([[G, G]]) \otimes F' \subset [\mathscr{L}(G), \mathscr{L}(G)] \otimes F'$, whence $\mathcal{L}([[G, G]]) \subset [\mathcal{L}(G), \mathcal{L}(G)]$. On the other hand, since $G/[[G, G]]$ is abelian, so is its Lie algebra, whence $[\mathcal{L}(G), \mathcal{L}(G)] \subset \mathcal{L}([[G, G]])$. This completes the proof of Theorem 13.3.

The following corollary is now almost evident.

COROLLARY 13.4. *Let G be an affine algebraic group over a field of characteristic 0. Then, for every Lie subalgebra L of $\mathcal{L}(G)$, the commutator subalgebra $[L, L]$ is an algebraic Lie subalgebra of $\mathcal{L}(G)$.*

Proof: By Proposition 13.1, we have $[L, L] = [[L], [L]]$, and by Theorem 13.3 we have $[[L], [L]] = \mathcal{L}([[G_L, G_L]])$.

Let G_u denote the unipotent radical of an affine algebraic group G over a field F of characteristic 0. Note that, by Theorem 10.1, G_u is connected, and $\mathcal{L}(G_u)$ is $\mathcal{A}(G)$-nilpotent. Since G_u is normal in G, and hence also in the connected component G_1 of the neutral element in G, we know from Theorem 9.5 that $\mathcal{L}(G_u)$ is an ideal of $\mathcal{L}(G)$. Conversely, if S is any $\mathcal{A}(G)$-nilpotent ideal of $\mathcal{L}(G)$, then (by Theorem 10.1) G_S is a unipotent algebraic subgroup of G_1, and is normal in G_1, by Theorem 9.5. Hence $G_S \subset (G_1)_u$. Clearly, $(G_1)_u$ is normal in G, so that $(G_1)_u \subset G_u$ (since G_u is connected, we have $G_u \subset (G_1)_u$, so that, actually, $G_u = (G_1)_u$). Thus we have $G_S \subset G_u$. Now it is clear that *$\mathcal{L}(G_u)$ is the unique largest $\mathcal{A}(G)$-nilpotent ideal of $\mathcal{L}(G)$*. We call it the *$\mathcal{A}(G)$-nilpotent radical of $\mathcal{L}(G)$*.

THEOREM 13.5. *Let (G, A) be the structure of an affine algebraic group over a field F of characteristic 0. Let N denote the A-nilpotent radical of $\mathcal{L}(G)$. Then there is an A-semisimple algebraic Lie subalgebra T of $\mathcal{L}(G)$ such that $\mathcal{L}(G)$ is the semidirect sum $N + T$.*

Proof: Let R denote the ordinary Lie algebra radical of $\mathcal{L}(G)$. By Levi's theorem, $\mathcal{L}(G)$ is a semidirect sum $R + S$, where S is a semisimple Lie subalgebra of $\mathcal{L}(G)$. Note that S is A-semisimple. Among the Lie subalgebras of $\mathcal{L}(G)$ that contain S and are A-semisimple, choose a maximal one, T say. From the remark preceding Proposition 13.1, we see that the algebraic hull $[T]$ of T in $\mathcal{L}(G)$ is still A-semisimple. Since T is maximal, this shows that T is an algebraic Lie subalgebra of $\mathcal{L}(G)$. From the fact that T is A-semisimple, it follows that every polynomial representation of G_T is semisimple (Theorem 9.1 and remark (2) preceding it). In particular, the restriction to G_T of the adjoint representation of G on $\mathcal{L}(G)$ is semisimple. Hence (Theorem 9.1) $\mathcal{L}(G)$ is semisimple also as a T-module. Since N and R are ideals of $\mathscr{L}(G)$, they are T-submodules. Evidently, $N \subset R$. Hence N has a T-module complement, P say, in R. As we have already noted in proving Theorem 13.3, $[\mathscr{L}(G), R]$ is A-nilpotent, and therefore lies in N. In particular, this implies that $[T, P] = (0)$.

In order to conclude that $\mathscr{L}(G) = N + T$, it evidently suffices to show that $R = (T \cap R) + N$. If this is not the case, then there is an element σ in P that does not belong to $N + T$. Referring to Theorem 11.1, consider the nilpotent and semisimple components $\sigma^{(n)}$ and $\sigma^{(s)}$ of σ in $\mathscr{L}(G)$. It is easy to see that $F\sigma^{(n)} + N$ is still an A-nilpotent ideal of $\mathscr{L}(G)$, so that it must coincide with N. Thus $\sigma^{(n)}$ belongs to N, whence $\sigma^{(s)}$ cannot belong to T. On the other hand, since $[\sigma, T] = (0)$, σ^* commutes with every element of T^*. The same is therefore true for $(\sigma^*)^{(s)} = (\sigma^{(s)})^*$ (see Theorem 1.15). Hence, the fact that T and $\sigma^{(s)}$ are A-semisimple implies that $F\sigma^{(s)} + T$ is still an A-semisimple Lie subalgebra of $\mathscr{L}(G)$, which contradicts the maximality of T. Thus we have shown that $\mathscr{L}(G) = N + T$.

Finally, note that $G_{T \cap N}$ is contained in $G_T \cap G_N$, and that $G_T \cap G_N$ is trivial, because it is unipotent as well as A-semisimple. Thus $G_{T \cap N}$ is trivial, so that $T \cap N = (0)$. This completes the proof of Theorem 13.5.

EXERCISES

1. Let F be field of characteristic 0, and let L be the Lie algebra of an affine algebraic group over F. Show that the image of L under the adjoint representation of L is an algebraic Lie subalgebra of the Lie algebra of the group of all F-linear automorphisms of the F-space L. Use this result and Theorem 11.1 in order to show that the following Lie algebra L cannot be the Lie algebra of an affine algebraic group over F: L has a basis (x, y, z), and the Lie composition is given by $[x, y] = 0$, $[z, x] = x + y$, $[z, y] = y$.
2. Let S be a finite-dimensional, not necessarily associative, algebra over a field F of characteristic 0. With the help of Theorem 8.3, prove that the Lie algebra of all derivations of S is an algebraic Lie subalgebra of the Lie algebra of the group of all linear automorphisms of S; more precisely, that it is the Lie algebra of the group of all F-algebra automorphisms of S.

14. SEMIDIRECT PRODUCT DECOMPOSITION

We wish to obtain the main results concerning reductive subgroups of affine algebraic groups over a field of characteristic 0. Note (referring to the beginning of Section 9) that *if $\rho: G \to H$ is a morphism of affine algebraic groups then ρ sends every reductive subgroup of G onto a reductive subgroup of H, and every unipotent subgroup of G onto a unipotent subgroup of H.*

The following auxiliary result involves the group $[G, G_u]$, which is defined as the subgroup of G that is generated by the commutators $xyx^{-1}y^{-1}$, with x ranging over G, and y over the unipotent radical G_u of G.

LEMMA 14.1. *Let (G, A) be the structure of an affine algebraic group over a field F of characteristic 0. Suppose that there is a reductive subgroup P of G such that $G_uP = G$. Let Q be any reductive subgroup of G. Then there is an element t in $[G, G_u]$ such that $tQt^{-1} \subset P$.*

Proof: We make an induction on the dimension of G_u. If this dimension is 0, then G_u (being connected) is trivial, and there is nothing to prove. Otherwise, let N denote the center of G_u. From the fact that G_u is unipotent, it follows readily that N is non-trivial whenever G_u is non-trivial. Clearly, N is a normal algebraic subgroup of G. By Proposition 11.2, P is an algebraic subgroup of G, and G is the semidirect product of G_u and P, so that A may be identified with $A^P \otimes A^{G_u}$. The factor A^P may be identified with $\mathscr{A}(G_u)$, and the factor A^{G_u} may be identified with $\mathscr{A}(P)$. Consistently with this, A^N is identified with $(A^P)^N \otimes A^{G_u}$. By Theorem 10.3, N is a *properly* normal algebraic subgroup of G_u, in the sense of Section 11. Hence our identifications show that N is also a properly normal algebraic subgroup of G, and that G/N is the semidirect product of G_u/N and the canonical image of P in G/N (which, simultaneously, is seen to be an algebraic subgroup of G/N, isomorphic with P). The image of Q in G/N is evidently a reductive subgroup of G/N. Clearly, G_u/N is the unipotent radical of G/N.

Now suppose that Lemma 14.1 has already been proved in the case where the unipotent radical has smaller dimension than the present G_u. Then we conclude that there is an element s in $[G, G_u]$ such that $sQs^{-1} \subset NP$. Therefore, we may now suppose that $Q \subset NP$. Let $\pi : G \to P$ be the morphism that is associated with our semidirect product decomposition $G = G_u \cdot P$. Define the polynomial map $\pi' : G \to G_u$ by $\pi'(x) = x\pi(x)^{-1}$. Then π' maps Q into N, and if x and y are elements of Q we have

$$\pi'(xy) = xy\pi(xy)^{-1} = xy\pi(y)^{-1}\pi(x)^{-1} = \pi'(x)\pi(x)\pi'(y)\pi(x)^{-1}.$$

For every x in Q, let $f(x) = \log(\pi'(x))$, so that f is a map of Q into $\mathscr{L}(N)$. Since log and π' are polynomial maps, f is the restriction to Q of a polynomial map of G into $\mathscr{L}(G_u)$. Since $\mathscr{L}(N)$ is abelian, the map $\exp : \mathscr{L}(N) \to N$ is a group homomorphism, and therefore so is its inverse $\log : N \to \mathscr{L}(N)$. Our above expression for $\pi'(xy)$ therefore gives

$$f(xy) = f(x) + \log(\pi(x)\pi'(y)\pi(x)^{-1}).$$

From Section 9 (the formula immediately preceding Theorem 9.3), we see that

$$\log(\pi(x)\pi'(y)\pi(x)^{-1}) = \alpha(x)(f(y)),$$

where α denotes the adjoint representation of G on $\mathscr{L}(G)$. For x in Q, and u in $\mathscr{L}(N)$, let us now abbreviate $\alpha(x)(u)$ by $x \cdot u$, and let us note that this defines a polynomial representation of Q on $\mathscr{L}(N)$, in the sense that the associated representative functions belong to A_Q. Our result now reads

$$f(xy) = f(x) + x \cdot f(y).$$

This identity shows that we obtain a Q-module structure on the direct sum $F + \mathscr{L}(N)$ if we define

$$x \cdot (r, u) = (r, rf(x) + x \cdot u)$$

for all x in Q, r in F, and u in $\mathscr{L}(N)$. Evidently, this is still a polynomial representation of Q. Since Q is reductive, it follows that our Q-module $F + \mathscr{L}(N)$ is semisimple. Hence the Q-submodule $\mathscr{L}(N)$ has a Q-module complement in $F + \mathscr{L}(N)$. This complement contains one and only one element of the form $(1, u)$. Since

$$x \cdot (1, u) = (1, f(x) + x \cdot u),$$

it follows that we must have $u = f(x) + x \cdot u$. Moreover, since the Q-module $\mathscr{L}(N)$ is semisimple, the element u is the sum of a Q-fixed element and an element v that is a sum of elements of the form $w - y \cdot w$, with w in $\mathscr{L}(N)$ and y in Q. The above gives $v = f(x) + x \cdot v$. We have

$$\exp(w - y \cdot w) = \exp(w)\exp(-y \cdot w) = \exp(w) y \exp(w)^{-1} y^{-1},$$

which belongs to $[G, G_u]$. Hence we conclude that $\exp(v)$ belongs to $[G, G_u]$. For every x in Q, we have

$$\begin{aligned} x\pi(x)^{-1} = \pi'(x) &= \exp(f(x)) \\ &= \exp(x \cdot v)^{-1}\exp(v) = x\exp(v)^{-1}x^{-1}\exp(v). \end{aligned}$$

Hence $\pi(x) = \exp(v)^{-1}x\exp(v)$, which shows that $\exp(v)^{-1}Q\exp(v)$ is contained in P. This completes the proof of Lemma 14.1.

THEOREM 14.2. *Let (G, A) be the structure of an affine algebraic group over a field F of characteristic 0. There is a reductive algebraic subgroup P of G such that G is the semidirect product of G_u and P. If Q is any reductive subgroup of G, then there is an element t in $[G, G_u]$ such that $tQt^{-1} \subset P$.*

Proof: In view of Proposition 11.2 and Lemma 14.1, all that remains to be proved is that there is a reductive subgroup P of G such that $G = G_uP$. First, we prove this in the case where G is connected. By Theorem 13.5, there is an A-semisimple algebraic Lie subalgebra T of $\mathscr{L}(G)$ such that $\mathscr{L}(G) = \mathscr{L}(G_u) + T$. Clearly, G_T is an A-semisimple algebraic subgroup of G. Consider the algebraic hull $[G_uG_T]$ of G_uG_T in G. Its Lie algebra evidently contains $\mathscr{L}(G_u)$ and T, so that it coincides with $\mathscr{L}(G)$. Assuming that G is connected, we have therefore from Proposition 8.5 that $[G_uG_T] = G$. We shall show that $[G_uG_T] = G_uG_T$. In order to do this, let F' be an algebraic closure of F, and consider the extended groups $G^{F'}$, $(G_u)^{F'}$, and $(G_T)^{F'}$. We know that the last two groups may be identified with the algebraic hulls in $G^{F'}$ of G_u and G_T, respectively. By Corollary 5.4, $(G_u)^{F'}(G_T)^{F'}$ is an algebraic subgroup of $G^{F'}$. Since it contains G_uG_T, which is algebraically dense in $G^{F'}$, we have therefore $(G_u)^{F'}(G_T)^{F'} = G^{F'}$. Now let Σ denote the Galois group of F' over F. We let Σ act on $A \otimes F'$ in the natural way through the factor F', and we identify the elements of Σ with the corresponding F-algebra automorphisms of $A \otimes F'$. It is then clear that an element x of $G^{F'}$ belongs to G if and only if $\sigma \circ x \circ \sigma^{-1} = x$ for every σ in Σ. Let x be an element of G, and write $x = yz$, with y in $(G_u)^{F'}$ and z in $(G_T)^{F'}$. Then we have $yz = (\sigma \circ y \circ \sigma^{-1})(\sigma \circ z \circ \sigma^{-1})$, whence $y^{-1}(\sigma \circ y \circ \sigma^{-1}) = z(\sigma \circ z \circ \sigma^{-1})^{-1}$. The expression on the left belongs to $(G_u)^{F'}$, while the expression on the right belongs to $(G_T)^{F'}$. Thus our element belongs to $(G_u)^{F'} \cap (G_T)^{F'}$. But $(G_u)^{F'}$ is $A \otimes F'$-unipotent, and (by Proposition 1.14) $(G_T)^{F'}$ is $A \otimes F'$-semisimple. Hence $(G_u)^{F'} \cap (G_T)^{F'}$ is trivial, so that we must have $y = \sigma \circ y \circ \sigma^{-1}$, and $z = \sigma \circ z \circ \sigma^{-1}$. Hence y belongs to $(G_u)^{F'} \cap G = G_u$, and z belongs to $(G_T)^{F'} \cap G = G_T$. Thus we have shown that $G_u G_T = G$, which establishes Theorem 14.2 in the case where G is connected.

Now we prove the general case by induction on the dimension of G_u. If G_u is trivial, then it is clear from what we have just shown, in view of Theorem 4.4, that the connected component G_1 of the neutral element in G is A-semisimple. Since G/G_1 is finite, it follows therefore from Proposition 12.1 that G is A-semisimple. Now suppose that G_u is non-trivial, and that Theorem 14.2 has been proved in the cases of lower dimensional unipotent radical. Let N be the center of G_u. In proving Lemma 14.1, we saw that N is a properly normal algebraic subgroup of G_1. By Theorem 4.4, N is therefore also a properly normal algebraic subgroup of G. By the inductive hypothesis, there is a reductive algebraic subgroup L of G/N such that G/N is the semidirect product of its unipotent radical G_u/N and L. By the connected case of Theorem 14.2, there is a reductive algebraic subgroup P_1 of G_1 such that G_1 is the semidirect product of G_u and P_1. The canonical image of P_1 in G/N is a reductive subgroup of G/N. By Lemma 14.1, it is contained in a conjugate of L. Hence we may choose L so that it contains the image of P_1 in G/N. Now we have

$$(G_u/N)L_1 = (G/N)_1 = G_1/N = (G_u/N)\big((NP_1)/N\big),$$

whence we see that L_1 coincides with the canonical image $(NP_1)/N$ of P_1. Moreover, it is clear from the semidirect product decompositions that the canonical map $P_1 \to L_1$ is an isomorphism of affine algebraic groups. Now let M be the inverse image of L in G, so that $M/N = L$, and $M_1/N = L_1$. We have $M_1 = NP_1$, by what we have seen just above. Let us make a coset decomposition $L = \bigcup_{i=1}^n x_i L_1$. For each i, choose an element y_i in M whose canonical image in L is x_i. We choose x_1 and y_1 to be the neutral elements of L and M, respectively. Then our isomorphism $L_1 \to P_1$ (the inverse of the above) extends to a map $\rho : L \to M$ such that $\rho(x_i u) = y_i \rho(u)$ for every i, and every u in L_1. It is clear from Theorem

4.4 that ρ is a polynomial map. The composite of ρ with the canonical map $M \to L$ is evidently the identity map on L.

Since N is abelian, we obtain an L-module structure on N by defining the transform of an element v of N by an element x of L to be $x \cdot v = \rho(x)v\rho(x)^{-1}$. Let us define the map $f: L \times L \to N$ by

$$f(x, y) = \rho(x)\rho(y)\rho(xy)^{-1}.$$

Writing this in the form $\rho(x)\rho(y) = f(x, y)\rho(xy)$, we see from associativity that

$$\big(x \cdot f(y, z)\big)f(x, yz) = f(x, y)f(xy, z)$$

for all triples (x, y, z) of elements of L. Since ρ is a polynomial map, so is f. For each element x of L, let f_x denote the polynomial map of L into N given by $f_x(y) = f(x, y)$. Via the group isomorphism $\log : N \to \mathscr{L}(N)$, we may regard N as an F-space, and then we may view f_x as an element of $N \otimes (A^N)_L$, noting that $(A^N)_L = \mathscr{A}(L)$. The above identity for f may be written in the form

$$\big(x \cdot f_y(z)\big)f_x(yz) = f(x, y)f_{xy}(z).$$

Viewing N as an F-space, and identifying the elements of N with the corresponding constant maps $L \to N$, we may write this as a relation among maps $L \to N$ in the additive form

$$x \cdot f_y + f_x \cdot y = f(x, y) + f_{xy},$$

where, for any map h of L into N, the left and right L-transforms of h are defined by $(x \cdot h)(z) = x \cdot h(z)$, and $(h \cdot y)(z) = h(yz)$. The corresponding actions of L on $N \otimes (A^N)_L$ are such that $x \cdot (v \otimes a) = (x \cdot v) \otimes a$, and $(v \otimes a) \cdot y = v \otimes (a \cdot y)$. Since L is reductive, $(A^N)_L$ is semisimple with respect to the action of L by right translations. Hence there is a right L-module projection $\pi : (A^N)_L \to F$. Define the map $g : L \to N$ by $g(x) = (i_N \otimes \pi)(f_x)$. Then g is clearly a polynomial map, and if we apply $i_N \otimes \pi$ to our above identity, we obtain

$$x \cdot g(y) + g(x) = f(x, y) + g(xy).$$

Reverting to the multiplicative notation, we have

$$\big(x \cdot g(y)g(x)\big) = f(x, y)g(xy)$$

for all elements x and y of L. This shows that, if h is the map of L into M defined by $h(x) = g(x)^{-1}\rho(x)$, then h is a group homomorphism. Clearly, the composite of h with the canonical map $M \to L$ is the identity map on L. Moreover, h is evidently a polynomial map, and thus is a morphism of affine algebraic groups $L \to M$. Now we have $M = Nh(L)$, and $h(L)$ is a reductive subgroup of M. Therefore, $h(L)$ is also a reductive subgroup of G. Since $(G_u/N)L = G/N$, we have $G_uM = G$, i.e., $G_uh(L) = G$. This completes the proof of Theorem 14.2.

We close this section with the following simple application of Theorem 14.2.

THEOREM 14.3. *Let $\rho: G \to H$ be a morphism of affine algebraic groups over a field of characteristic* 0. *Suppose that $\rho(G)$ is algebraically dense in H. Then $\rho(G_u) = H_u$.*

Proof: Clearly, $\rho(G_u)$ is a normal unipotent subgroup of H, so that $\rho(G_u) \subset H_u$. Let L denote the inverse image of H_u in G. Then L is a normal algebraic subgroup of G containing G_u, whence $L_u = G_u$. By Theorem 14.2, there is a reductive subgroup P of L such that $L = G_uP$. Now $\rho(P)$ is an $\mathscr{A}(H)$-semisimple subgroup of H. Since $\rho(P) \subset H_u$, it follows that $\rho(P)$ is trivial, so that $\rho(L) = \rho(G_u)$. Since $\rho(G)$ is algebraically dense in H, we know from Theorem 7.5 that $\rho^\circ(\mathscr{L}(G)) = \mathscr{L}(H)$. The inverse image of $\mathscr{L}(H_u)$ in $\mathscr{L}(G)$ is $\mathscr{L}(L)$, and $\rho^\circ(\mathscr{L}(L)) = \rho^\circ(\mathscr{L}(G_u))$, because $\rho(L) = \rho(G_u)$. Hence we have $\rho^\circ(\mathscr{L}(G_u)) = \mathscr{L}(H_u)$. Using Theorem 10.1, we now obtain

$$H_u = \exp(\mathscr{L}(H_u)) = \exp(\rho^\circ(\mathscr{L}(G_u))) = \rho(\exp(\mathscr{L}(G_u))) = \rho(G_u),$$

so that Theorem 14.3 is proved.

EXERCISES

1. In the notation of Theorem 14.2, note that $[G, G_u]$ is a properly normal algebraic subgroup of G, and show that the factor group $G/[G, G_u]$ is isomorphic, as an affine algebraic group, with the direct product $(G_u/[G, G_u]) \times P$. Conclude from this that $[G, G_u]P$ is an algebraic subgroup of G. Write $C_G{}^1(G_u)$ for $[G, G_u]$, and $C_G^{i+1}(G_u)$ for $[G, C_G{}^i(G_u)]$. Show that, for large enough i, one has $C_G{}^i(G_u) = C_G^{i+1}(G_u)$, and call this limit group $C_G{}^\infty(G_u)$. Now prove that the element t of Theorem 14.2 may actually be taken from $C_G{}^\infty(G_u)$.

2. Let V be a finite-dimensional vector space over an arbitrary field F, and let G be the group of all linear automorphisms of V. Show that the unipotent radical G_u of G is trivial (using the fact that there must be a non-zero element of V that is fixed under the action of G_u). Now refer to Exercise 1 of Section 12 in order to show that the assumption

that F be of characteristic 0 is needed for the validity of Theorem 14.2.

3. Let F be an algebraically closed field of characteristic 0. Determine all connected affine algebraic groups over F whose dimension does not exceed 2.

15. AUTOMORPHISM GROUPS

Let (G, A) be the structure of an affine algebraic group over an arbitrary field F. We consider the group $\mathscr{W}(G)$ of all affine algebraic group automorphisms of G. In particular, we wish to examine the possibility of endowing $\mathscr{W}(G)$ with a natural affine algebraic group structure. A minimum requirement concerning the significance of such a structure is that the map $G \times \mathscr{W}(G) \to G$ sending each element (x, α) of $G \times \mathscr{W}(G)$ onto $\alpha(x)$ be a polynomial map. This amounts to requiring that, for every element f of A, and every element τ of the dual A° of A, the map $\tau/f\colon \mathscr{W}(G) \to F$, where $(\tau/f)(\alpha) = \tau(f \circ \alpha)$, be a polynomial function on $\mathscr{W}(G)$. Let us view A as a right $\mathscr{W}(G)$-module, with $\mathscr{W}(G)$ acting by composition $f \to f \circ \alpha$ on A. Clearly, a necessary condition for the existence of a suitable affine algebraic group structure on $\mathscr{W}(G)$ is that A be locally finite as a right $\mathscr{W}(G)$-module. If G satisfies this condition, then we say that G is *conservative*.

Assume that G is conservative, and write W for $\mathscr{W}(G)$. Define $\mathscr{A}(W)$ as the smallest fully stable algebra of F-valued functions on W containing the above functions τ/f. Then $\mathscr{A}(W)$ is clearly a Hopf algebra of representative functions on W.

THEOREM 15.1. *Let G be a conservative affine algebraic group over an arbitrary field F, and let W be the group of all affine algebraic group automorphisms of G. Then $(W, \mathscr{A}(W))$, with $\mathscr{A}(W)$ defined as above, is an affine algebraic group structure over F, and the natural map $G \times W \to G$ is a polynomial map.*

Proof: It is clear from the definition of $\mathscr{A}(W)$ that, as an F-algebra, $\mathscr{A}(W)$ is generated by the functions of the form τ/f and their antipodes $(\tau/f)'$, where $(\tau/f)'(\alpha) = (\tau/f)(\alpha^{-1})$. Let us write A for $\mathscr{A}(G)$. The assumption that G is conservative implies that, for every element f of

A, there are F-linearly independent elements $f_1, \ldots, f_n$ in A such that $f \circ \alpha = \sum_{i=1}^{n} g_i(\alpha) f_i$, where the g_i's are uniquely determined F-valued functions on W. The space spanned by these g_i's is precisely the space of functions τ/f, with τ ranging over A°. Since A is finitely generated as an F-algebra, this shows that the same is true for $\mathscr{A}(W)$. Clearly, $\mathscr{A}(W)$ separates the elements of W. What remains to be proved is that every F-algebra homomorphism $\mathscr{A}(W) \to F$ is the evaluation at an element of W.

Let σ be an F-algebra homomorphism $\mathscr{A}(W) \to F$. For every element x of G, define the map $\sigma_x : A \to F$ by $\sigma_x(f) = \sigma(x/f)$. Clearly, σ_x is an F-algebra homomorphism, i.e., σ_x is an element of G. If x and y are elements of G, and if α is an element of W, we have

$$\big((xy)/f\big)(\alpha) = (xy)(f \circ \alpha) = f\big(\alpha(x)\alpha(y)\big) = \big(\alpha(x) \otimes \alpha(y)\big)\big(\gamma(f)\big),$$

where γ is the comultiplication of A. This shows that

$$\sigma_{xy} = (\sigma_x \otimes \sigma_y) \circ \gamma = \sigma_x \sigma_y .$$

Thus the map $x \to \sigma_x$ is a group homomorphism $G \to G$. Denoting this homomorphism by σ', we have

$$(f \circ \sigma')(x) = \sigma_x(f) = \sigma(x/f)$$

for every f in A and every x in G. From the beginning of this proof, we know that, for any given element f of A, the functions x/f, with x ranging over G, all lie in a finite-dimensional F-subspace of $\mathscr{A}(W)$. Hence the restriction of σ to the functions x/f is a finite linear combination of evaluations at elements of W, i.e., there are elements $c_1, \ldots, c_n$ of F, and elements $\alpha_1, \ldots, \alpha_n$ of W, such that $\sigma(x/f) = \sum_{i=1}^{n} c_i x(f \circ \alpha_i)$ for every x in G. Thus $f \circ \sigma'$ is the element $\sum_{i=1}^{n} c_i f \circ \alpha_i$ of A, and we have shown that σ' is a morphism of affine algebraic groups. Moreover, if π denotes the antipode of $\mathscr{A}(W)$, one verifies directly from the definitions that $(\sigma \circ \pi)'$ is the inverse of σ'. Thus σ' is an automorphism of affine algebraic groups, i.e., σ' is an element of W. Finally, from the fact that $(x/f)(\sigma') = \sigma(x/f)$ for every x in G and every f in A, one sees that σ coincides with the evaluation of $\mathscr{A}(W)$ at σ', because the functions x/f, together with their antipodes, generate $\mathscr{A}(W)$ as an F-algebra. Thus we have shown that $\big(W, \mathscr{A}(W)\big)$ is the structure of an affine algebraic group. After this, the last part of Theorem 15.1 follows at once from the definition of $\mathscr{A}(W)$, so that Theorem 15.1 is proved.

PROPOSITION 15.2. *Let G be a conservative affine algebraic group over an arbitrary field F. Let L be the Lie algebra of G, and let $\mathscr{W}(L)$ denote* *the affine algebraic group of all Lie algebra automorphisms of L. Then the natural map $\mathscr{W}(G) \rightarrow \mathscr{W}(L)$ is a morphism of affine algebraic groups.*

Proof. Let us denote the natural map $\mathscr{W}(G) \rightarrow \mathscr{W}(L)$ by τ. This means that, for every element α of $\mathscr{W}(G)$, $\tau(\alpha)$ is the differential of α, so that $\tau(\alpha)(\rho)(f) = \rho(f \circ \alpha)$ for every element ρ of L, and every element f of $\mathscr{A}(G)$.

Let $\mathscr{E}(L)$ denote the F-space of all F-linear endomorphisms of L. The restrictions to $\mathscr{W}(L)$ of the elements of the dual $\mathscr{E}(L)^\circ$ of $\mathscr{E}(L)$, together with the reciprocal of the determinant function on $\mathscr{W}(L)$, generate the algebra of polynomial functions of the affine algebraic group $\mathscr{W}(L)$. For every element f of $\mathscr{A}(G)$, and every element ρ of L, let f/ρ denote the element of $\mathscr{E}(L)^\circ$ that is defined by $(f/\rho)(e) = e(\rho)(f)$. Then the linear functions f/ρ evidently span $\mathscr{E}(L)^\circ$ over F. Hence, in order to prove that τ is a morphism of affine algebraic groups, it suffices to show that, for each f/ρ as above, the composite $(f/\rho) \circ \tau\colon \mathscr{W}(G) \rightarrow F$ belongs to $\mathscr{A}(\mathscr{W}(G))$. For every α in $\mathscr{W}(G)$, we have

$$(f/\rho)(\tau(\alpha)) = \tau(\alpha)(\rho)(f) = \rho(f \circ \alpha)\,.$$

Hence $(f/\rho) \circ \tau$ is simply the element ρ/f of $\mathscr{A}(\mathscr{W}(G))$, as defined at the beginning of this section. Proposition 15.2 is therefore established.

It is easy to see that, *if the connected component G_1 of the neutral element in an affine algebraic group G is conservative, then G is conservative. The converse is false.* From now on, we shall confine our attention to connected affine algebraic groups over algebraically closed fields of characteristic 0. First, we show that certain special types of groups are conservative.

Let F be an algebraically closed field of characteristic 0, and let S be a connected *semisimple* affine algebraic group over F, in the sense that $\mathscr{L}(S)$ is a semisimple Lie algebra. We shall show that S is conservative. Let Q denote the group of all Lie algebra automorphisms of $\mathscr{L}(S)$, and consider the adjoint representation $\alpha : S \rightarrow Q$. We know from Section 9 that the differential of α sends $\mathscr{L}(S)$ onto the Lie algebra of all *inner derivations* D_x of $\mathscr{L}(S)$, where $D_x(y) = [x, y]$. Since $\mathscr{L}(S)$ is a semisimple Lie algebra, every derivation of $\mathscr{L}(S)$ is an inner derivation. Hence the differential of α is an isomorphism of $\mathscr{L}(S)$ onto the Lie algebra of all derivations of $\mathscr{L}(S)$ (note that the center of $\mathscr{L}(S)$ is (0)). If G is the full linear group $\mathscr{G}(\mathscr{L}(S))$, then $\mathscr{L}(G)$ is the Lie algebra $\mathscr{E}(\mathscr{L}(S))$ of all linear endomorphisms of $\mathscr{L}(S)$. It is easy to see from the definition of the groups G_σ in Section 8 that, if σ is an element of

$\mathscr{E}(\mathscr{L}(S))$, then G_σ is contained in Q if and only if σ is a derivation of $\mathscr{L}(S)$. Hence $\mathscr{L}(Q)$ is precisely the Lie algebra of all derivations of $\mathscr{L}(S)$, so that the differential of α is an isomorphism of $\mathscr{L}(S)$ onto $\mathscr{L}(Q)$. Since F is algebraically closed, $\alpha(S)$ is a connected algebraic subgroup of Q. Hence it is clear that $\alpha(S)$ coincides with the connected component of the neutral element in Q. In particular, $\alpha(S)$ is of finite index in Q.

Since F is of characteristic 0, the natural map $\tau: \mathscr{W}(S) \to Q$ is injective, as is seen from Corollary 9.2. Let S' denote the group of the inner automorphisms of S. Then S' is a normal subgroup of $\mathscr{W}(S)$. By the definitions of τ and α, we have $\tau(S') = \alpha(S)$. By the above, $\tau(S')$ is therefore of finite index in $\tau(\mathscr{W}(S))$ (actually, in Q), so that S' is of finite index in $\mathscr{W}(S)$. Evidently, $\mathscr{A}(S)$ is locally finite as an S'-module. Since S' is normal and of finite index in $\mathscr{W}(S)$, this implies that $\mathscr{A}(S)$ is locally finite as a $\mathscr{W}(S)$-module, so that S is conservative.

Now let us consider an affine algebraic group G over F that is a semidirect product $G_u \cdot S$, where S is a connected semisimple algebraic subgroup of G. We shall show that G is conservative. Let X denote the stabilizer of S in $\mathscr{W}(G)$. Since S is a maximal reductive subgroup of G, it follows at once from the conjugacy part of Theorem 14.2 that $XG' = \mathscr{W}(G)$, where G' is the group of inner automorphisms of G. The restriction image X_S of X in $\mathscr{W}((S)$ contains S', and we know from the above that X_S/S' is finite. It follows that, if Y is the element-wise fixer of S in $\mathscr{W}(G)$, then YS' is of finite index in $XS' = \mathscr{W}(G)$. Moreover, YS' is normal in $\mathscr{W}(G)$. Hence, in order to show that G is conservative, it suffices to prove that $\mathscr{A}(G)$ is locally finite as a YS'-module, and hence it suffices to prove that $\mathscr{A}(G)$ is locally finite as a Y-module.

Write A for $\mathscr{A}(G)$, and note that we have a tensor product decomposition $A = A^S \otimes A^{G_u}$, corresponding to the semidirect product decomposition $G = G_u \cdot S$. Clearly, Y acts trivially on A^{G_u}, and stabilizes A^S. If A^S is identified with $\mathscr{A}(G_u)$ by the restriction map, then the action of Y on A^S becomes the transpose of the action of Y on G_u. For the unipotent group G_u, the map $f \to f \circ \exp$ is an F-algebra isomorphism of $\mathscr{A}(G_u)$ onto the algebra $\mathscr{A}(\mathscr{L}(G_u))$ of polynomial functions on the F-space $\mathscr{L}(G_u)$. If α is an element of $\mathscr{W}(G_u)$, then $\alpha \circ \exp = \exp \circ \tau(\alpha)$, by Theorem 10.2. Evidently, the action on $\mathscr{A}(\mathscr{L}(G_u))$ of the group of Lie algebra automorphisms of $\mathscr{L}(G_u)$ is locally finite. It follows that $\mathscr{A}(G_u)$ is locally finite as a $\mathscr{W}(G_u)$-module. In particular, A^S is therefore locally finite as a Y-module, and the above shows that A is therefore locally finite as a Y-module. Thus we have shown that G is conservative.

The following lemma is an extension of the result we have just obtained.

LEMMA 15.3. *Let G be a connected affine algebraic group over the algebraically closed field F of characteristic* 0. *Let P be a maximal reductive subgroup of G, and let T be the connected component of the neutral element in the center of P. Then G is conservative if and only if the restriction image in $\mathscr{W}(T)$ of the stabilizer of T in $\mathscr{W}(G)$ is finite.*

Proof: We note that T is a toroid (by Theorem 11.4), so that $\mathscr{A}(T) = F[f_1, \ldots, f_n, f_1^{-1}, \ldots, f_n^{-1}]$, where the f_i's are algebraically independent homorphisms of T into the multiplicative group F^* of F. Let α be any element of $\mathscr{W}(T)$. Then each $f_i \circ \alpha$ is a morphism $T \to F^*$, and consequently (cf. proof of Theorem 11.4) is a product of integral (positive or negative) powers of the f_i's. If α_{ij} is the exponent of f_j in $f_i \circ \alpha$, then the map that sends each α onto the matrix with entries α_{ij} is an isomorphism of $\mathscr{W}(T)$ cnto the group of integral n by n matrices with determinant 1 or -1. As α ranges over an infinite subset of $\mathscr{W}(T)$, there is no upper bound for the absolute values of the exponents α_{ij}, whence we see that $\mathscr{A}(T)$ cannot be locally finite as a module for any infinite subgroup of $\mathscr{W}(T)$. Since $\mathscr{A}(T)$ is the restriction image of $\mathscr{A}(G)$, the assumption that G is conservative implies that the action on $\mathscr{A}(T)$ of the restriction image in $\mathscr{W}(T)$ of the stabilizer of T in $\mathscr{W}(G)$ is locally finite. Therefore, the condition of Lemma 15.3 is necessary.

Now suppose that the condition of Lemma 15.3 is satisfied. Let X denote the stabilizer of P in $\mathscr{W}(G)$. As before, we see from the conjugacy part of Theorem 14.2 that $XG' = \mathscr{W}(G)$. Clearly, X stabilizes T. Let Z denote the element-wise fixer of T in $\mathscr{W}(G)$. Our assumption implies that $Z \cap X$ is of finite index in X. Hence ZG' is of finite index in XG', i. e., in $\mathscr{W}(G)$. Moreover, Z is stable under the conjugation action of X, and is conjugated into ZG' by the elements of G', so that ZG' is normal in $\mathscr{W}(G)$. Therefore, in order to show that G is conservative, it suffices to show that $\mathscr{A}(G)$ is locally finite as a Z-module.

Since P is reductive, we see readily from the structure theory of Lie algebras, as used in the proof of Theorem 12.2, that $P = ST$, where S is a connected semisimple algebraic subgroup of P, and in fact is the commutator subgroup of P. Also, by Theorem, 14.2, we have the semidirect product decomposition $G = G_u \cdot P$. Write H for the algebraic subgroup $G_u \cdot S$ of G. One sees readily that $H = [G, G]G_u$, whence it is clear that H is $\mathscr{W}(G)$-stable. Combining the comultiplication of $\mathscr{A}(G)$ with the map $\mathscr{A}(G) \otimes \mathscr{A}(G) \to \mathscr{A}(H) \otimes \mathscr{A}(T)$ obtained from the restriction maps, we have an F-algebra homomorphism σ: $\mathscr{A}(G) \to \mathscr{A}(H) \otimes \mathscr{A}(T)$. From the fact that $G = HT$, we see that σ is injective. Since Z stabilizes H, we have a right Z-module structure on $\mathscr{A}(H)$, via the restriction homomorphism $Z \to \mathscr{W}(H)$. We extend this to a Z-module structure on $\mathscr{A}(H) \otimes \mathscr{A}(T)$ under which the elements of $\mathscr{A}(T)$ are Z-fixed. Since Z leaves the elements of T fixed, the above map σ is then a homomorphism of Z-modules. Now we have $H = G_u \cdot S$, whence it is clear that $H_u = G_u$, so that H is of the type discussed above Lemma 15.3. Hence H is conservative. In particular, $\mathscr{A}(H)$, and therefore also $\mathscr{A}(H) \otimes \mathscr{A}(T)$, is locally finite as a Z-module. Since σ is an injective Z-module homomorphism, this implies that $\mathscr{A}(G)$ is locally finite as a Z-module, so that Lemma 15.3 is proved.

THEOREM 15.4. *Let G be a connected affine algebraic group over the algebraically closed field F of characteristic* 0. *Let C denote the center of G. Then G is conservative if and only if one of the following two conditions is satisfied:* (1) *C/C_u is finite;* (2) *the dimension of the center of G/G_u is at most* 1.

Proof: By Theorem 14.2, there is a reductive subgroup P of G such that G is the semidirect product of G_u and P. Let T be the connected component of the neutral element in the center of P. Since P is isomorphic with G/G_u, condition (2) is equivalent to the condition that the dimension of T be at most 1. Hence, if condition (2) is satisfied, then $\mathscr{W}(T)$ is finite (trivial, or of order 2), and it follows at once from Lemma 15.3 that G is conservative.

Now suppose that condition (1) is satisfied, and let Y denote the stabilizer of T in $\mathscr{W}(G)$. By Lemma 15.3, G is conservative if the restriction image of Y in $\mathscr{W}(T)$ is finite. Clearly, Y stabilizes the algebraic subgroup G_uT of G. Hence, in order to show that G is conservative, it suffices to show that, if Q is the stabilizer of T in $\mathscr{W}(G_uT)$, the restriction image of Q in $\mathscr{W}(T)$ is finite. In order to do this, consider the adjoint action of T on $\mathscr{L}(G_u)$. Since T is a toroid, we can decompose $\mathscr{L}(G_u)$ into a direct sum $V_{f_1} + \cdots + V_{f_n}$ of T-submodules, where the f_i's are mutually distinct morphisms $T \to F^*$, and $t \cdot v = f_i(t)v$ for every t in T and every v in V_{f_i} ($t \cdot v$ denotes the t-transform of v under the adjoint representation). If α is an element of Q, and α_T is its restriction image in $\mathscr{W}(T)$, then the automorphism $\tau(\alpha)$ of $\mathscr{L}(G_u)$ that corresponds to α maps each V_{f_i} onto some V_{f_j}, with $f_j = f_i \circ \alpha_T^{-1}$. In this way, we obtain a homomorphism δ of Q into the finite group of permutations of the set $(f_1, \ldots, f_n)$. Now let α be an element of the kernel of δ. Then the adjoint action of $\alpha(t)t^{-1}$ on $\mathscr{L}(G_u)$ is trivial for every element t of T. Hence each $\alpha(t)t^{-1}$ centralizes G_u, and therefore lies in the center C of G. Let σ be the map of T into C/C_u that sends each element t of T onto the canonical image of $\alpha(t)t^{-1}$ in C/C_u. Then σ is a morphism of affine algebraic groups $T \rightarrow C/C_u$. By our assumption that C/C_u is finite, σ must therefore be the trivial map, because $\sigma(T)$ is a connected algebraic subgroup of C/C_u (Corollary 5.2). This means that each $\alpha(t)t^{-1}$ lies in C_u. The map $t \rightarrow \alpha(t)t^{-1}$ is therefore a morphism of affine algebraic groups $T \rightarrow C_u$. Since T is reductive, while C_u is unipotent, we must therefore have $t = \alpha(t)$. Thus we have shown that the restriction image in $\mathscr{W}(T)$ of the kernel of δ is trivial. Since the kernel of δ is of finite index in Q, this shows that the restriction image of Q in $\mathscr{W}(T)$ is finite, so that G is conservative.

It remains to be shown that if neither (1) nor (2) is satisfied then G is not conservative. Appealing to Corollary 11.6, let us write $T = T_0 \times T_1$, where T_0 is the connected component of the neutral element in $C \cap T$, and T_1 is a complementary toroid. First, consider the case where T_0 is of dimension greater than 1. As in the proof of Lemma 15.3, we have $P = ST$, where S is a semisimple algebraic subgroup of P. Hence $G = (G_u ST_1)T_0$. The factor $G_u ST_1$ is an algebraic subgroup of G (by Corollary 5.4), and its intersection with T_0 is finite. Hence, as is easy to see, there are infinitely many elements of $\mathscr{W}(T_0)$ that leave the elements of $(G_u ST_1) \cap T_0$ fixed. Clearly, each of these extends to an element of $\mathscr{W}(G)$ that leaves the elements of $G_u ST_1$ fixed. Now observe that T_0 cannot be trivial, because otherwise condition (1) would be satisfied. Hence, if the dimension of T_0 is not greater than 1, then T_0 must be 1-dimensional. Since condition (2) is assumed not to be satisfied either, T_1 is then non-trivial. Therefore, there are then infinitely many morphisms $\rho: T_1 \rightarrow T_0$ whose kernels contain the finite group $(G_u ST_0) \cap T_1$. For each such ρ, we have an element ρ^* of $\mathscr{W}(G)$ such that ρ^* leaves the elements of $G_u ST_0$ fixed, while $\rho^*(t) = t\rho(t)$ for every element t of T_1. Thus, in any case, the restriction image in $\mathscr{W}(T)$ of the stabilizer of T in $\mathscr{W}(G)$ is infinite. By Lemma 15.3, G is therefore not conservative, so that Theorem 15.4 is proved.

THEOREM 15.5. *Let G be a connected affine algebraic group over the algebraically closed field F of characteristic 0. Then G is conservative if and only if the canonical image of $\mathscr{W}(G)$ in $\mathscr{W}(\mathscr{L}(G))$ is an algebraic subgroup of $\mathscr{W}(\mathscr{L}(G))$.*

Proof: First, suppose that G is conservative. Then we know from Proposition 15.2 that the natural map $\mathscr{W}(G) \to \mathscr{W}(\mathscr{L}(G))$ is a morphism of affine algebraic groups. Since F is algebraically closed, the image of $\mathscr{W}(G)$ in $\mathscr{W}(\mathscr{L}(G))$ is therefore an algebraic subgroup of $\mathscr{W}(\mathscr{L}(G))$ (it contains the image of $\mathscr{W}(G)_1$ as a subgroup of finite index, and the image of $\mathscr{W}(G)_1$ is an algebraic subgroup of $\mathscr{W}(\mathscr{L}(G))$, by Corollary 5.2).

Now suppose that the image of $\mathscr{W}(G)$ is an algebraic subgroup of $\mathscr{W}(\mathscr{L}(G))$. Let τ denote the canonical map $\mathscr{W}(G) \to \mathscr{W}(\mathscr{L}(G))$. Let T be a toroid as in Lemma 15.3, and let X be the stabilizer of T in $\mathscr{W}(G)$. Proceeding as in the proof of Theorem 9.5, we see that $\tau(X)$ is precisely the stabilizer of $\mathscr{L}(T)$ in $\tau(\mathscr{W}(G))$, whence $\tau(X)$ is still an algebraic subgroup of $\mathscr{W}(\mathscr{L}(G))$. Hence the restriction image of $\tau(X)$ in $\mathscr{W}(\mathscr{L}(T))$ is an algebraic subgroup of $\mathscr{W}(\mathscr{L}(T))$, and it is evidently contained in the canonical image of $\mathscr{W}(G)$. In fact, it is the canonical image of the restriction image of X in $\mathscr{W}(T)$. By Lemma 15.3, it suffices to show that the restriction image of X in $\mathscr{W}(T)$ is finite. Since the canonical map $\mathscr{W}(T) \to \mathscr{W}(\mathscr{L}(T))$ is injective, it suffices to show that the restriction image of $\tau(X)$ in $\mathscr{W}(\mathscr{L}(T))$ is finite.

In order to do this, observe that $\mathscr{W}(\mathscr{L}(T))$ is simply the group $\mathscr{G}(\mathscr{L}(T))$ of all linear automorphisms of the F-space $\mathscr{L}(T)$. From the beginning of our proof of Lemma 15.3, we see that there is an F-basis of $\mathscr{L}(T)$ with respect to which the canonical image of $\mathscr{W}(T)$ appears as the group of all integral matrices of determinant 1 or -1. Now let H be any algebraic subgroup of $\mathscr{W}(\mathscr{L}(T))$ that lies in the canonical image of $\mathscr{W}(T)$. Let f_{ij}, with i and j ranging from 1 to the dimension n of T, be the polynomial functions on $\mathscr{W}(\mathscr{L}(T))$ such that $f_{ij}(\alpha)$ is the (i, j)-entry of the matrix representing α for every α in $\mathscr{W}(\mathscr{L}(T))$. Let g_{ij} be the restriction image of f_{ij} in $\mathscr{A}(H_1)$. If H_1 is non-trivial, then at least one of the g_{ij}'s is non-constant, and is therefore not algebraic over F, because F is algebraically closed and $\mathscr{A}(H_1)$ is an integral domain. Let u be such a g_{ij}. By Theorem 1.3, there is a non-zero polynomial $p(u)$ in $F[u]$ such that every F-algebra homomorphism $F[u] \to F$ not annihilating $p(u)$ is the evaluation at some element h of

H_1. Clearly, there are such F-algebra homomorphisms $F[u] \to F$ that map u onto an element of F that is not a rational integer. On the other hand, $h(u)$ is a rational integer, because H_1 consists of integral matrices. Thus we have a contradiction. Our conclusion is that H_1 is trivial, so that H is a finite group. In particular, the restriction image of $\tau(X)$ in $\mathscr{W}(\mathscr{L}(T))$ is finite, so that our proof of Theorem 15.5 is now complete.

We shall apply Theorem 15.4 in order to show that the automorphism group of a solvable Lie algebra is always conservative. However, we require some preliminary information, which is provided by the following proposition.

PROPOSITION 15.6. *Let L be a finite-dimensional Lie algebra over a perfect field F, and suppose that there is a non-trivial semisimple automorphism α of L that commutes with every derivation of L. Let L^α denote the α-fixed part of L. Then L is the direct Lie algebra sum $T + L^\alpha$, with T abelian, and $[L^\alpha, L^\alpha] = L^\alpha$.*

Proof: Let F' be an algebraic closure of F, and let α' be the automorphism of $L \otimes F'$ that is induced by α. Since F is perfect, we know from Proposition 1.14 that α' is still semisimple. Now it is readily seen that if the conclusions of Proposition 15.6 hold for $(L \otimes F', \alpha')$ then they hold also for (L, α). Hence we may assume without loss of generality that F is algebraically closed. Assuming this, let $L = \sum_a L_a$ be a characteristic subspace decomposition of L with respect to α. Here, the index a ranges over a certain finite set of non-zero elements of F, and α acts as the scalar multiplication by a on each L_a. We have $L^\alpha = L_1$, and we put $T = \sum_{a \neq 1} L_a$. Evidently, L is then the direct F-space sum $T + L^\alpha$. Let x be an element of an L_a, with $a \neq 1$. From the fact that α commutes with the inner derivation effected by x, we find that $\alpha(x) - x$ lies in the center, Z say, of L. Thus $ax - x$ belongs to Z. Since $a \neq 1$, this shows that x belongs to Z. Hence we have $T \subset Z$.

It follows that $[L, L] = [L^\alpha, L^\alpha] \subset L^\alpha$. Suppose that $L^\alpha \neq [L, L]$, and choose an element x of L^α that does not belong to $[L, L]$. Since α is non-trivial, there is an $a \neq 1$ such that $L_a \neq (0)$. Let z be a non-zero element of such an L_a. There is a linear map $\rho: L \to Z$ such that $\rho([L, L]) = (0)$ and $\rho(x) = z$. Clearly, ρ is a derivation of L, and therefore commutes with α. In particular, $\rho(\alpha(x)) = \alpha(\rho(x))$, i.e., $z = az$, with $a \neq 1$, contradicting the assumption that $z \neq 0$. Hence we must have $L^\alpha = [L, L] = [L^\alpha, L^\alpha]$, which proves Proposition 15.6.

COROLLARY 15.7. *Let L be a finite-dimensional non-abelian solvable Lie algebra over a perfect field F. Then the centralizer of the connected component of the neutral element in the automorphism group of L is unipotent.*

Proof: Let $\mathscr{W}(L)$ denote the group of all Lie algebra automorphisms of L. The Lie algebra of $\mathscr{W}(L)$ is the Lie algebra $\mathscr{D}(L)$ of all derivations of L, as we have seen in our discussion preceding Lemma 15.3. Let σ denote the representation of $\mathscr{W}(L)$ on the algebra $\mathscr{E}(L)$ of all linear endomorphisms of L given by $\sigma(\alpha)(h) = \alpha \circ h \circ \alpha^{-1}$. The differential σ° of this polynomial representation σ sends every element δ of $\mathscr{D}(L)$ onto the endomorphism $h \rightarrow \delta \circ h - h \circ \delta$ of $\mathscr{E}(L)$. In order to see this, note that $\mathscr{E}(L)$ may be identified with the Lie algebra of $\mathscr{G}(L)$, and that our representation σ is then the restriction to $\mathscr{W}(L)$ of the adjoint representation of $\mathscr{G}(L)$ (see the formula immediately preceding Theorem 9.3). Hence the differential of σ is the restriction to $\mathscr{G}(L)$ of the adjoint representation of $\mathscr{E}(L)$, by the general statement immediately preceding Theorem 9.4.

An element of the centralizer of $\mathscr{W}(L)_1$ is fixed under $\sigma(\mathscr{W}(L)_1)$, and hence is annihilated by $\sigma^\circ(\mathscr{D}(L))$, i.e., it commutes with every derivation of L. Since L is solvable and non-abelian, it is clear from Proposition 15.6 that the centralizer of $\mathscr{W}(L)_1$ can therefore contain no non-trivial semisimple element. By Theorem 11.1, it follows that every element of this centralizer is unipotent. By Corollary 10.7, this implies the conclusion of Corollary 15.7.

THEOREM 15.8. *Let L be a solvable Lie algebra over an algebraically closed field F of characteristic* 0. *Then the automorphism group of L is conservative.*

Proof: First, consider the case where L is abelian. In this case, $\mathscr{W}(L)$ is the group $\mathscr{G}(L)$ of all linear automorphisms of L. It is clear from Theorem 12.2 that $\mathscr{G}(L)$ is a reductive affine algebraic group, and it is evidently connected. Moreover, the center of $\mathscr{G}(L)$ is 1-dimensional (isomorphic with F^*). Thus $\mathscr{G}(L)$, i.e., $\mathscr{W}(L)$, satisfies condition (2) of Theorem 15.4, and is therefore conservative.

Now suppose that L is non-abelian. Then we have from Corollary 15.7 that the center of $\mathscr{W}(L)_1$ is unipotent. Thus $\mathscr{W}(L)_1$ satisfies condition (1) of Theorem 15.4, so that $\mathscr{W}(L)_1$ is conservative. As we have already remarked at the beginning of this section, it follows easily from this (using Theorem 4.4) that $\mathscr{W}(L)$ is conservative. This establishes Theorem 15.8.

COROLLARY 15.9. *Let G be a unipotent affine algebraic group over an algebraically closed field of characteristic* 0. *Then the automorphism group* $\mathscr{W}(G)$ *is conservative.*

Proof: It is clear from Theorems 10.1 and 10.2 that the canonical map $\mathscr{W}(G) \to \mathscr{W}(\mathscr{L}(G))$ is an isomorphism of affine algebraic groups whenever G is unipotent. Hence Corollary 15.9 follows from Theorem 15.8.

We remark that Corollary 15.9 does not extend to the case where G is solvable. A simple example to show this is the following. Let F^+ denote the additive group of F, viewed as a unipotent affine algebraic group in the natural way. Let $F^+ \cdot F^*$ denote the usual semidirect product of F^+ and the multiplicative group F^* of F, in which the conjugation action of F^* on F^+ is by scalar multiplication. Finally, let G be the direct product $F^+ \times (F^+ \cdot F^*)$. One can show by direct computation that $\mathscr{W}(G)$ is then isomorphic with the direct product $F^* \times (F^+ \cdot F^*)$, which satisfies neither one of the conditions of Theoren 15.4, so that $\mathscr{W}(G)$ is not conservative.

Theorem 15.8 *holds also for semisimple Lie algebras L.* If L is semisimple, then the Lie algebra of the center of $\mathscr{W}(L)_1$ is (0), owing to the fact that the center of L is (0), and that all derivations of L are inner. Hence $\mathscr{W}(L)_1$ satisfies condition (1) of Theorem 15.4, so that $\mathscr{W}(L)_1$ is conservative, whence $\mathscr{W}(L)$ is conservative. However, *Theorem* 15.8 *does not hold for arbitrary Lie algebras L.* The following is a counter-example. Let S be a semisimple Lie algebra such that $\mathscr{W}(S)$ is connected. For instance, one may take S to be the Lie algebra of all endomorphisms of trace 0 of a 2-dimensional F-space. Let U be a simple non-trivial S-module, and let K be the semidirect Lie algebra sum $U + S$, with U viewed as an abelian ideal. Let F stand for the 1-dimensional Lie algebra, and let L be the direct Lie algebra sum of F and K. Then it can be shown that $\mathscr{W}(L)$ is not conservative.

EXERCISES

1. Let F be an algebraically closed field of characteristic 0. Let G_1 be the 2-dimensional toroid $F^* \times F^*$. Let σ be the automorphism of G_1 given by $\sigma(x, y) = (y, x)$, and let G be the semidirect product of G_1 by the group of order 2 that is generated by σ, so that in G, we have $\sigma u = \sigma(u)\sigma$, for every element u of G_1. Show that $\mathscr{W}(G)/G'$ is finite, where G' is the group of inner automorphisms of G. Hence G is conservative, while the component G_1 of the neutral element in G is not conservative.

2. Verify the assertions made in the text concerning the properties of the example immediately following Corollary 15.9.
3. Let G and H be affine algebraic groups over an arbitary field F. Suppose there is given a group homomorphism $\alpha : H \to \mathscr{W}(G)$ such that $\mathscr{A}(G)$ is locally finite with respect to the right H-module structure obtained from α, and the associated representative functions on H belong to $\mathscr{A}(H)$. Define the semidirect product $G \cdot H$ so that $(x_1, y_1)(x_2, y_2) = (x_1 a(y_1)(x_2), y_1 y_2)$. Show that $G \cdot H$ may be made into an affine algebraic group such that G and H are algebraic subgroups of $G \cdot H$, and $G \cdot H$ is their semidirect product in the sense of affine algebraic groups.

SUPPLEMENTARY READING

1. A. Borel, *Linear Algebraic Groups* (notes taken by H. Bass), Benjamin, New York, 1969
2. C. Chevalley, *Theorie des groupes de Lie*, vols. II (1951) and III (1955), Hermann, Paris
3. ———, *Séminaire sur la classification des groupes de Lie algebriques*, École Normale Superieure (mimeographed), Paris, 1956–58
4. G. Hochschild, *The Structure of Lie Groups*, Holden-Day, San Francisco, 1965
5. N. Jacobson, *Lie Algebras*, Interscience, New York, 1962
6. M. E. Sweedler, *Hopf Algebras*, Benjamin, New York, 1969

INDEX